RADIATION HEAT TRANSFER NOTES

D. K. EDWARDS
University of California, Los Angeles

HEMISPHERE PUBLISHING CORPORATION
Washington New York London

RADIATION HEAT TRANSFER NOTES

1 2 3 4 5 6 7 8 9 0 B C B C 8 9 8 7 6 5 4 3 2 1

Library of Congress Cataloging in Publication Data

Edwards, D. K. (Donald Kenneth), date
Radiation heat transfer notes.

Includes bibliographies and index.
1. Heat—Radiation and absorption. I. Title.
TJ260.E318 621.402'2 81-4539
ISBN 0-89116-231-3 AACR2

CONTENTS

Preface *vii*
Principal Symbols *ix*

1 INTRODUCTION *1*

1.A Radiation heat transfer in thermal design *1*
1.B Thermodynamic surfaces and surface systems *2*
1.C Radiant intensity and flux *5*
1.D Black-body radiation *8*
1.E Fractional functions *11*
References *16*
Exercises *17*

2 SURFACE RADIATION CHARACTERISTICS *19*

2.A Introduction to surface characteristics *19*
2.B Absorption and emission characteristics *20*
2.C Measurement of absorption and emission characteristics *25*
2.D Reflection and transmission characteristics *29*
2.E Measurement of reflection and transmission characteristics *33*
2.F Electromagnetic theory and the Fresnel relations *36*
2.G Some useful approximations *46*
2.H Polarization *54*
2.I Thermal radiation characteristics in thermal design *61*
References *80*
Exercises *86*

3 RADIATION TRANSFER BETWEEN PERFECTLY DIFFUSE SURFACES *93*

3.A Shape Factors *93*
3.B Radiosity-irradiation formulations *107*
3.C Refractory surfaces *116*
3.D The radiation network *117*
3.E Selected working relations *123*
3.F Diffuse-walled passages *137*
References *146*
Exercises *149*

4 RADIATION TRANSFER BETWEEN NONDIFFUSE WALLS *157*

4.A Specular and imperfectly diffuse surfaces *157*
4.B The mirror image concept *158*
4.C The Monte Carlo algorithm *161*
4.D Specular-walled passages *167*
4.E Surface models *177*
4.F Rough-walled passages *184*
References *187*
Exercises *191*

5 GAS RADIATION PROPERTIES *195*

5.A The equation of transfer *195*
5.B Measurement of gas radiation properties *200*
5.C Qualitative remarks on the physics of gas radiation *206*
5.D Spectral, band, and total property definitions *209*
5.E Molecular gas radiation properties *220*
5.F Gas mixtures *232*
References *236*
Exercises *239*

6 RADIATION TRANSFER WITH AN ISOTHERMAL GAS *245*

6.A Heat transfer at a black wall *245*
6.B The mean beam length concept *246*
6.C Wall layer transmission *253*
6.D Radiosity-irradiation formulation at an isothermal gas-filled-enclosure wall *254*
6.E The radiation network with a gas *256*
6.F Some useful results *260*
6.G Sample calculation *266*
6.H Monte Carlo solution *269*
References *274*
Exercises *276*

7 NONISOTHERMAL GAS RADIATION *281*

7.A Solution of the equation of transfer *281*
7.B Geometrical considerations *286*
7.C The slab geometry *289*
7.D Differential formulations *294*
7.E Spectral considerations, scaling approximations *307*
7.F Molecular gas radiation in the slab *311*
References *321*
Exercises *323*

8 **RADIATION ACTING WITH CONDUCTION OR CONVECTION** *329*

8.A Combined phenomena *329*
8.B Thermal network analysis *330*
8.C Radiation-cooled transient heating or cooling *338*
8.D Radiant heat exchanger *339*
8.E The radiating fin *341*
8.F Linearized radiation and conduction around the hollow cylinder *346*
8.G Convection of a radiation molecular gas *353*
References *363*
Exercises *366*

PREFACE

These Notes are intended for a 40-lecture-hour course in radiation heat transfer. The emphasis is on developing the information and skills to formulate and solve engineering heat transfer problems. The Notes are in two main parts. After an introduction to the fundamental concepts of spectral and total intensity and flux and fractional functions, the first part develops the subject of surface-to-surface radiation heat transfer. Surface radiation characteristics needed to incorporate radiation heat transfer rates into a first law of thermodynamics heat balance are defined and related to fundamental physical properties where appropriate via electromagnetic theory. The experimental techniques developed to measure the characteristics are briefly described. Then the engineering enclosure problem is addressed. Radiosity-irradiation formulations, the network method, the mirror-image concept, and the powerful Monte Carlo algorithm are described. The second part considers the subject of radiation transfer in a participating medium. Gas radiation properties are defined, and their measurement described. The engineering enclosure problem with a well-stirred and thus homogeneous absorbing-emitting medium is treated using radiosity-irradiation and network representations. The Monte Carlo algorithm is extended to cover an absorbing-emitting-scattering medium that can be homogeneous or inhomogeneous. Geometrical and spectral complexities and simplifications are considered, primarily for the slab. Differential formulations are briefly described. Narrow- and wide-band scaling for inhomogeneous molecular gases complete the second part concerned with gas radiation. The Notes conclude with a short section on heat transfer by radiation and conduction or convection.

To keep within the confines of a 40-hour course, the Notes keep to the engineering radiation heat transfer discipline. Even so, there is an attempt to develop two levels of problem-solving skills. The first level, of great importance to practicing engineers, is what may be termed finite-element thermal system analysis. This skill is emphasized first in the development, in Sections 3 and 6, and in the early portion of Section 8. The second level is differential-element or so-called exact analysis. While this level is almost totally absent from Sections 3 and 6, the bulk of Sections 7 and 8 are devoted to it.

Grateful acknowledgment is made to Mrs. Phyllis Gilbert, who typed the Notes. Professor Frank W. Schmidt reviewed the text for incorporation into the *Heat Exchanger Design Handbook* and made several helpful suggestions. Messrs. J. C. McMurrin and W. A. Menard helped considerably in proofreading the typescript. Mr. Patrick Hubbard assisted with the drafting of many of the illustrations. The help of graduate students who participated in the radiation heat transfer course at UCLA over the past several years is also gratefully acknowledged.

D. K. Edwards

PRINCIPAL SYMBOLS

a	length between surface roughness asperities, μm
A, A_i, A_j	area, m^2
A_c	cross-sectional area, m^2
A_k	band absorption of kth band, cm^{-1}
A_k^*	band absorptance at kth band, dimensionless
A^+	Van Driest's constant, 26
b	thickness, m; also Mei and Squire constant, 3.4
B	black body heat flux, W/m^2, or the spectral black body heat flux, $W/m^2\ cm^{-1}$ or W/m^2 μm; also the rotational constant, cm^{-1}
$\underline{B}$	magnetic induction
c	velocity of light, 2.997925×10^8 m/s
c_1	first radiation constant, 3.7415×10^{-16} J m^2/s
c_2	second radiation constant, 1.4388 cm K
c_f	skin friction coefficient
c_p	specific heat at constant pressure, J/kg K
C*	channel emissivity, dimensionless
d	spectral line spacing, cm^{-1}
D	diameter, m; also optical collision diameter
$\underline{D}$	electric induction
E	energy, J or eV
E_n	exponential integral of order n
$\underline{E}$	electric field strength
f	function of fractional function; also friction factor
f_e	external fractional function, dimensionless

f_i	internal fractional function, dimensionless
F	collision frequency, s^{-1}; also entry length factor
F_{i-j}	shape factor from area i to area j, dimensionless
$\mathscr{F}_{i-j}$	transfer factor from area i to area j, dimensionless
g	statistical weight
G	normal irradiation of a beam, $\int I d\Omega$, W/m^2
G_{i-j}	area-shape-factor product, $A_i F_{i-j}$, m^2
$\mathscr{G}_{i-j}$	area-transfer-factor product $A_i \mathscr{F}_{i-j}$; also the area-to-volume or volume-to-volume quantity, m^2
h	Planck's constant, $6.6256x10^{-34}$ Js
h_c	convective heat transfer coefficient
h_r	radiative heat transfer coefficient
$\hat{h}$	specific enthalpy, J/kg
$\underline{H}$	magnetic field strength
I	intensity, W/m^2 sr; also the spectral intensity, W/m^2 cm^{-1} sr or W/m^2 μm sr
I_b	black body intensity (units as above)
$\underline{j}$	electric current density
k	absorptive index; also Boltzmann's constant, $1.38054x10^{-23}$ J/K
k_a	absorption coefficient, m^{-1}
k_e	extinction coefficient, m^{-1}
k_s	scattering coefficient, m^{-1}
K	thermal conductivity, W/m K; also the second moment of the radiant intensity, also von Karman's constant, 0.4
K_R	radiation conductivity, W/m K
K*	radiation kernel

L	length, m
L_{mb}	mean beam length, m
L_{mbg}	geometric mean beam length, m
m	mass, kg
$\dot{m}$	mass flow rate, kg/s
n	refractive index
N	radiation conductivity to molecular conductivity ratio, dimensionless
N_v	number density, m^{-3}
Nu	Nusselt number
P	pressure, N/m^2 or atm
$\mathcal{P}$	perimeter, m
P_e	dimensionless equivalent line broadening pressure
P_i	partial pressure of species i
Pr_m	Prandtl number
Pr_t	turbulent Prandtl number
q	heat flux, W/m^2
q^+	radiosity, W/m^2
q^-	irradiation, W/m^2
Q	heat energy, J
$\dot{Q}$	heat flow, W
Q_a	absorption efficiency
Q_e	extinction efficiency
Q_s	scattering efficiency

r_e electrical resistivity, ohm cm

r_{WL} wall-layer thickness-to-hydraulic diameter ratio, dimensionless

R radius, m

R_t turbulent Reynolds number

Re Reynolds number based upon hydraulic diameter

s slant path length, m

S integrated line intensity; also mean radiant intensity

t time, s; also optical depth, dimensionless

t_H optical depth at band head, dimensionless

T temperature, K

T_e environmental temperature or equivalent mean temperature, K

T_g gas temperature, K

T_r mean radiant temperature, K

T_w wall temperature, K

u sec θ; also cos θ; also velocity in x-direction, m/s

U unit step function; also average velocity in x-direction, m/s

v cos θ

V detector signal, volts

x length coordinate, m; also mole fraction

X density-path-length product, kg/m^2 or g/m^2

y length coordinate, m

z length coordinate, m

α absorptivity; also thermal diffusivity, m^2/s

α_k integrated band intensity, $cm^{-1}/(g/m^2)$

α_m molecular thermal diffusivity, m^2/s

β line-width-to-spacing parameter; also dimensionless temperature ratio

γ line half-width, cm^{-1}; also Euler's constant 0.5772156...

δ thickness or half thickness, m; also vibrational quantum number change

ε emissivity

ε_{eff} heat exchanger effectiveness

ε_H eddy diffusivity for heat

ε_M eddy diffusivity for momentum

η line-width-to-spacing parameter; also fin effectiveness

θ polar angle

κ mass absorption coefficient, $(kg/m^2)^{-1}$

λ wavelength, μm

μ cos θ

μ_m magnetic permeability

ν wavenumber, cm^{-1}; also kinematic viscosity, m^2/s

ν_f frequency, s^{-1}

ν_m molecular kinematic viscosity, m^2/s

π 3.1415926...

ρ reflectivity; also density kg/m^3

ρ_a absorber partial density, kg/m^3

ρ_e electric change density

σ Stefan-Boltzmann constant, 5.67×10^{-8} W/m^2 K^4

σ_e electrical conductivity

σ_r surface roughness, μm

τ transmissivity

$\tau_{H,k}$ optical depth at band head

τ_{WL} wall layer transmissivity

ϕ azimuthal angle

Φ hot band line spacing parameter, dimensionless

Ψ hot band intensity parameter, dimensionless

ω exponential band width, cm^{-1}

ω_s albedo for single scatter

Ω solid angle, sr

1 *INTRODUCTION*

1.A Radiation Heat Transfer in Thermal Design. When does one consider radiation heat transfer, and when does one not? One does not consider radiation inside of a fluid that is highly opaque to the source spectrum. In a fluid such as water, the radiation is merely a contributor to what we know as thermal conductivity. Similarly one does not consider radiation inside a fluid that is perfectly transparent to the source spectrum. If there is no physical mechanism by which the fluid can absorb energy from radiation passing through it, then it follows from thermodynamics that it cannot emit radiation either, and it cannot be either heated or cooled by radiation. Such a fluid is said to be diathermanous. The walls surrounding such a fluid, however, may exchange heat radiation, but only if they are not isothermal. Thus one does not ordinarily consider radiation within the passages of a heat exchanger containing oil, water, or air. The first two are opaque. The last is diathermanous.

When two walls at different temperatures are in view of one another or one wall is in view of a participating medium (one neither opaque nor diathermanous) the radiation heat flux (W/m^2) tends to be high when $\Delta\sigma T^4$ is high, where σ is the Stefan-Boltzmann constant 5.6697×10^{-8} $W/m^2 K^4$. When ΔT is small compared to the absolute temperature level, $\Delta\sigma T^4$ can be written $4\sigma T_m^3 \Delta T$ where T_m is the mean temperature level.

At 300 K the value for $4\sigma T_m^3$ is slightly over 6 W/m^2 K (circa 1 Btu/hr ft^2 R) on the same order as a natural convection heat transfer coefficient. At T_m = 2000 K the value is nearly 300 times greater. From such a value, 1800 W/m^2 K, one can see why radiation contributes to film boiling heat transfer. Radiation is important when temperatures are high, distances are large (because convective heat transfer coefficients go like passage size D as $D^{-1/5}$ for turbulent flow or D^{-1} for laminar flow), or under vacuum conditions when convective heat transfer coefficients are low because of the low fluid density.

1.B Thermodynamic Surfaces and Surface Systems. The thermal designer needs to know surface heat fluxes adjacent to the interface between phases. When one phase is highly opaque, and the other is not, the opaque surface system concept is used. Figure 1-1 depicts a surface system. The s-surface lies just outside the highly opaque phase; the u-surface lies just within it. The m-surface lies sufficiently below the phase interface so that (1) no radiation crossing the s and u surfaces is transmitted to the m-surface, and (2) the radiation flux crossing the m-surface is given by the radiation-diffusion equation and is included with the conduction. For no flow through the surfaces and negligible transient heat storage in the mass between the m and u surfaces, one has

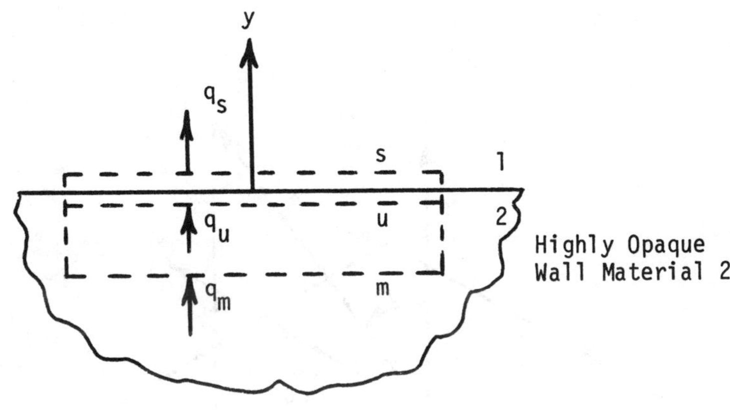

FIGURE 1-1

The s, u, and m Thermodynamic Boundaries
for a Surface System

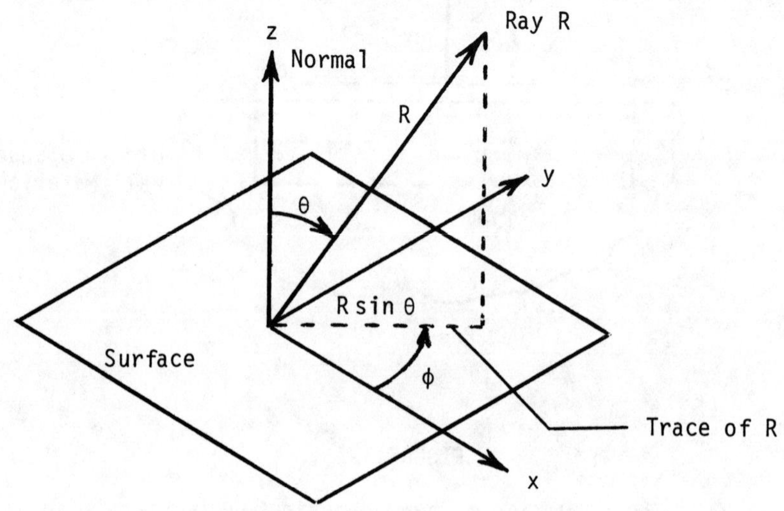

FIGURE 1-2

Polar Angle θ and Azimuthal Angle ϕ

$$q_s = q_u = q_m = -K_2 \left.\frac{dT_2}{dy}\right|_m \tag{1-1}$$

where q denotes heat flux (W/m^2). In the more transparent medium

$$q_s = -K_1 \left.\frac{dT_1}{dy}\right|_s + q_r \tag{1-2}$$

where q_r is the net radiant flux across the s-surface in medium 1. Note that there is no need to define a radiation heat flux in the opaque medium 2.

1.C Radiant Intensity and Flux. The radiant intensity describes the directional distribution of the radiant flux at an s-surface. The radiant flux itself may be separated into two parts, an outgoing flux q^+ away from medium 2 and an incoming flux q^- toward medium 2. Other symbols used for the radiosity q^+ are J and B and for the irradiation q^-, G or H. The net flux is

$$q_r = q = q^+ - q^- \tag{1-3}$$

The outward flux per unit solid angle is the intensity I^+. Solid angle is defined by area dA_s on a sphere of radius R

$$d^2\Omega \equiv \frac{dA_s}{R^2} \tag{1-4}$$

In polar-azimuthal coordinates illustrated in Fig. 1-2, dA_s is $Rd\theta R\sin\theta d\phi$ so that

$$d^2\Omega = \sin\theta d\theta d\phi \tag{1-5}$$

The unitless solid angle is attributed units of steradians, just as a plane angle is said to be measured in radians. The flux per unit solid angle I^+ is power per unit area per unit solid angle W/m^2 sr, but the normal of the area is understood to be in the direction of I^+. The contribution of $I^+d^2\Omega$ to the flux across the s-surface is

$$d^2q^+ = I^+\cos\theta d^2\Omega \tag{1-6}$$

Similarly

$$d^2q^- = I^-\cos\theta d^2\Omega \tag{1-7}$$

Combining Eqs. (1-5) and (1-6)

$$q^+ = \int_0^{2\pi}\int_0^{\pi/2} I^+(\theta,\phi)\cos\theta\sin\theta d\theta d\phi \tag{1-8}$$

Similarly

$$q^- = \int_0^{2\pi}\int_0^{\pi/2} I^-(\theta,\phi)\cos\theta\sin\theta d\theta d\phi \tag{1-9}$$

When $I^+(\theta,\phi)$ is independent of direction the radiation is said to be diffuse. For diffuse radiation

$$q^+ = \pi I^+ \qquad I^+ \neq f(\theta,\phi) \tag{1-10}$$

$$q^- = \pi I^- \qquad I^- \neq f(\theta,\phi) \tag{1-11}$$

For heat exchangers where long cylindrical tubes are often employed, Figure 1-3 shows base-axial angular coordinates. In these coordinates $dA_s = Rd\beta R\cos\beta d\gamma$ and

$$d^2\Omega = \cos\beta d\beta d\gamma \tag{1-12}$$

$$\cos\theta = \cos\beta\cos\gamma \tag{1-13}$$

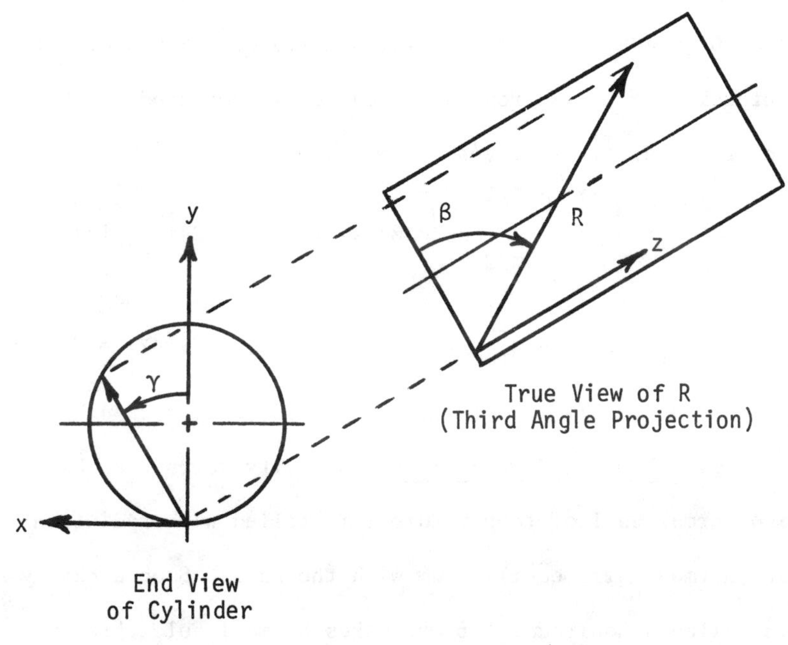

FIGURE 1-3

Base Angle γ and Axial Angle β

$$d^2q^{\pm} = I^{\pm}\cos^2\beta d\beta \cos\gamma d\gamma \qquad (1\text{-}14)$$

$$q = \int_{\gamma=-\pi/2}^{+\pi/2}\int_{\beta=-\pi/2}^{+\pi/2} I^{\pm}(\beta,\gamma)\cos^2\beta d\beta\cos\gamma d\gamma \qquad (1\text{-}15)$$

For example, what is the irradiation upon the closest element of wall parallel to an infinitely-long 2.5 cm diameter rod 10 cm away? The rod radiates diffusely 1850 W/m. If the intensity of the surrounds is negligible compared to that of the rod, Eq. (1-15) gives

$$q^- = I^-\int_{-\gamma_1}^{+\gamma_1}\int_{-\pi/2}^{+\pi/2}\cos^2\beta d\beta\cos\gamma d\gamma = [I^-][2\sin\gamma_1][\pi/2]$$

$$I^-_{wall} = I^+_{rod} = q^+_{rod}/\pi = [\dot{Q}/\pi DL][1/\pi] \;,\; 2\sin\gamma_1 = 2.5/10$$

$$q^- = [1850/\pi(.025)(1)][1/\pi][2.5/10][\pi/2] = 2944 \text{ W/m}^2$$

1.D Black-Body Radiation. A cavity surrounded by an isothermal wall of temperature T is filled with radiation in thermodynamic equilibrium with the wall. Such a cavity is called a hohlraum. If one makes a small hole through the cavity wall the equilibrium is disturbed only slightly. The radiation observed escaping from the hohlraum is black-body radiation. The word black is used because the hohlraum absorbs all radiation entering the hole in the wall and reflects none, so it appears black when it is cold. The intensity of radiation I_b^+ leaving the hohlraum is isotropic; black-body radiation is diffuse. Stefan's Law relates the black-body intensity and radiosity to the fourth power of the temperature

$$q_b^+ = \pi I_b^+ = \sigma T^4 \quad (1\text{-}16)$$

where the Stefan-Boltzmann constant σ is

$$\sigma = 5.6697\text{x}10^{-8} \text{ W/m}^2 \text{ K}^4$$

The radiation packets emitted by the random thermal agitations of electrically charged particles in the hohlraum walls travel at the speed of light c as a collapsing electric field builds up a magnetic field that in turn collapses to build up an electric field further along the path of propagation. The energy E, frequency ν_f, wavenumber ν, and wavelength λ are interrelated through the Einstein photoelectric law

$$E = h\nu_f = hc\nu = hc/\lambda \quad (1\text{-}17)$$

where h is Planck's constant and c is the velocity of light

$$h = 6.6256\text{x}10^{-34} \text{ J s}, \quad c = 2.997925\text{x}10^{8} \text{ m/s}$$

The relations between wavenumber ν, frequency ν_f, and wavelength λ are

$$\nu = \nu_f/c = 1/\lambda \quad (1\text{-}18)$$

In dimensional terms

$$\nu[\text{cm}^{-1}] = 10^4/\lambda[\mu\text{m}], \qquad \lambda[\mu\text{m}] = 10^4/\nu[\text{cm}^{-1}]$$

$$1[\mu\text{m}] = 10^{-6}[\text{m}] = 10^4[\text{Å}] \qquad (25.4[\mu\text{m}] = 10^{-3}[\text{inches}])$$

Very high energy photons may be described in terms of their energy in electron volts

$$E[\text{eV}] = 1.240/\lambda[\mu\text{m}]$$

Spectral intensity and spectral flux are respectively intensity and flux per unit band width of the spectrum

$$I_{b\nu} = \frac{dI_b}{d\nu} , \quad I_{b\lambda} = \frac{dI_b}{d\lambda} \tag{1-19}$$

$$q^+_{b\nu} = \frac{dq^+_b}{d\nu} , \quad q^+_{b\lambda} = \frac{dq^+_b}{d\lambda} \tag{1-20}$$

$$q^+_{b\nu} d\nu = q^+_{b\lambda} d\lambda , \quad d\lambda = (1/\nu^2) d\nu$$

$$q^+_{b\lambda} = \nu^2 q^+_{b\nu} , \quad q^+_{b\lambda} = (1/\lambda^2) q^+_{b\nu} \tag{1-21}$$

$$d^3q^{\pm} = I^{\pm}_{\nu} \cos\theta d^2\Omega d\nu \tag{1-22}$$

$$q^{\pm} = \int_0^{\infty} q^{\pm}_{\nu} d\nu = \int_0^{\infty} q^{\pm}_{\lambda} d\lambda \tag{1-23}$$

$$I^{\pm} = \int_0^{\infty} I^{\pm}_{\nu} d\nu = \int_0^{\infty} I^{\pm}_{\lambda} d\lambda \tag{1-24}$$

Planck's Law gives the spectral distribution of black-body radiation

$$I_{b\nu} = \frac{2hc^2\nu^3}{e^{hc\nu/kT}-1} , \quad I_{b\lambda} = \frac{2hc^2\lambda^{-5}}{e^{hc/\lambda kT}-1} \tag{1-25}$$

$$q^+_{b\nu} = \frac{2\pi hc^2\nu^3}{e^{hc\nu/kT}-1} , \quad q^+_{b\lambda} = \frac{2\pi hc^2\lambda^{-5}}{e^{hc/\lambda kT}-1} \tag{1-26}$$

where k is Boltzmann's constant, the gas constant divided by Avogadro's number

$$k = 1.38054\text{x}10^{-23} \text{ J/K}$$

The first and second radiation constants are

$$c_1 = 2\pi hc^2 = 3.7415\text{x}10^{-16} \text{ J m}^2 \text{ s}^{-1}$$

$$c_2 = hc/k = 1.4388\text{x}10^{-2} \text{ m K}$$

$$= 1.4388 \text{ cm K} = 14388 \ \mu\text{m K}$$

The Stefan-Boltzmann constant is given by

$$q_b^+ = \int_0^\infty \frac{2\pi hc^2\nu^3}{e^{hc\nu/kT}-1} d\nu = \sigma T^4$$

$$\sigma = \frac{2\pi k^4}{h^3c^2}\int_0^\infty \frac{\zeta^3 d\zeta}{e^\zeta - 1} = \frac{2\pi^5 k^4}{15h^3c^2} \qquad (1\text{-}27)$$

Two convenient limits are the Rayleigh Jeans limit (large T/ν)

$$e^{hc\nu/kT}-1 \doteq \frac{hc\nu}{kT} = \frac{hc}{\lambda kT} \qquad (1\text{-}28)$$

and the Wien limit (small T/ν)

$$e^{hc\nu/kT}-1 \doteq e^{hc\nu/kT} \qquad (1\text{-}29)$$

In the Rayleigh-Jeans limit $q_{b\lambda}^+$ goes as T. In the Wien limit $q_{b\lambda}^+$ goes like $e^{-c_2/\lambda T}$, like an Arrhenius chemical reaction rate. As the temperature increases, more black-body radiation is radiated at every wavelength or wavenumber, but, *relatively*, more in the short wavelengths and less in the long wavelengths.

1.E Fractional Functions. The fraction of the total black-body radiosity between wavelengths 0 and λ is called the fractional function of the first kind or the external fraction f_e

$$f_e(\lambda,T) = \frac{1}{\sigma T^4}\int_0^\lambda q_b^+(\lambda,T)d\lambda$$

$$f_e(\lambda T) = \frac{15}{\pi^4}\int_{\zeta}^{\infty}\frac{\zeta^3 d\zeta}{e^{\zeta}-1}, \quad \zeta = \frac{c_2}{\lambda T} \tag{1-30}$$

Wien's displacement law is that $\partial q_b^+(\lambda,T)/\partial\lambda = 0$ occurs at a fixed value of λT

$$\frac{\partial q_b^+}{\partial \lambda} = 0 \quad \text{at} \quad \lambda T = \frac{c_2}{4.965} = 2898\ \mu\text{m K}$$

This value of λT corresponds to a value of $f_e = 0.25$. But the value has no special sanctity, for on a wavenumber basis

$$\frac{\partial q_b^+}{\partial \nu} = 0 \quad \text{at} \quad \lambda T = \frac{c_2}{2.821} = 5100\ \mu\text{m K}$$

A fractional function of the second kind or the internal fraction f_i is defined by

$$f_i(\lambda T) \equiv \frac{1}{4\sigma T^3}\int_0^{\lambda}[\partial q_b^+(\lambda,T)/\partial T]d\lambda \tag{1-31}$$

Table 1-1 gives λT values for various values of f_e or f_i. Also shown are Δf values for numerical integration based upon the trapezoidal rule. Table 1-2 gives λ values for a few nonblack-body radiation sources.

For example, consider material being dried by a nearly-black electric heater element 10 cm away that is 2.5 cm in diameter and radiates 1850 W/m. The material absorbs 95% of the irradiation with wavelengths longer than 2.7 μm but only 30% of that with shorter wavelengths. What is the absorbed flux? Recall that in the previous example with the same geometry we found $q^- = 2944$ W/m^2. From Stefan's Law

TABLE 1-1

VALUES OF λT AT SPECIFIED VALUES OF THE EXTERNAL AND INTERNAL FRACTIONS OF BLACKBODY RADIATION

f	Δf	$\Sigma\Delta f$	$(\lambda T)_e$ cm K	$(\lambda T)_i$ cm K
0.0025	0.005	0.005	0.1230	0.1073
0.0075	0.005	0.010	0.1395	0.1209
0.0125	0.005	0.015	0.1495	0.1291
0.0175	0.005	0.02	0.1573	0.1352
0.025	0.01	0.03	0.1662	0.1423
0.035	0.01	0.04	0.1762	0.1501
0.045	0.01	0.05	0.1848	0.1565
0.055	0.01	0.06	0.1922	0.1625
0.07	0.02	0.08	0.202	0.1701
0.09	0.02	0.10	0.214	0.1791
0.11	0.02	0.12	0.225	0.1874
0.13	0.02	0.14	0.235	0.1949
0.15	0.02	0.16	0.245	0.202
0.17	0.02	0.18	0.254	0.209
0.19	0.02	0.20	0.263	0.216
0.21	0.02	0.22	0.272	0.222
0.23	0.02	0.24	0.281	0.229
0.25	0.02	0.26	0.290	0.235
0.27	0.02	0.28	0.299	0.242
0.29	0.02	0.30	0.308	0.248
0.31	0.02	0.32	0.317	0.255
0.33	0.02	0.34	0.326	0.261
0.35	0.02	0.36	0.335	0.268
0.37	0.02	0.38	0.344	0.275
0.39	0.02	0.40	0.353	0.281
0.41	0.02	0.42	0.363	0.288
0.43	0.02	0.44	0.373	0.295
0.45	0.02	0.46	0.384	0.303
0.47	0.02	0.48	0.394	0.310
0.49	0.02	0.50	0.405	0.318
0.51	0.02	0.52	0.416	0.326
0.53	0.02	0.54	0.428	0.334
0.55	0.02	0.56	0.440	0.342
0.57	0.02	0.58	0.453	0.351
0.59	0.02	0.60	0.467	0.361
0.61	0.02	0.62	0.482	0.371
0.63	0.02	0.64	0.497	0.382
0.65	0.02	0.66	0.513	0.393
0.67	0.02	0.68	0.531	0.405
0.69	0.02	0.70	0.550	0.417
0.71	0.02	0.72	0.570	0.430

TABLE 1-1 (continued)

f	Δf	ΣΔf	$(\lambda T)_e$ cm K	$(\lambda T)_i$ cm K
0.73	0.02	0.74	0.593	0.446
0.75	0.02	0.76	0.616	0.462
0.77	0.02	0.78	0.642	0.479
0.79	0.02	0.80	0.671	0.498
0.81	0.02	0.82	0.704	0.521
0.83	0.02	0.84	0.741	0.546
0.85	0.02	0.86	0.783	0.574
0.87	0.02	0.88	0.825	0.609
0.89	0.02	0.90	0.899	0.653
0.91	0.02	0.92	0.982	0.690
0.93	0.02	0.94	1.090	0.774
0.945	0.01	0.95	1.200	0.846
0.955	0.01	0.96	1.298	0.909
0.965	0.01	0.97	1.433	0.997
0.975	0.01	0.98	1.63	1.123
0.9825	0.005	0.985	1.87	1.28
0.9875	0.005	0.990	2.22	1.43
0.9925	0.005	0.995	2.56	1.70
0.9975	0.005	1.000	7.34	2.49

TABLE 1-2

VALUES OF WAVELENGTH AT SPECIFIED VALUES OF THE EXTERNAL FRACTION

f	Δf	ΣΔf	Wavelength λ, μm			
			Extraterrestrial Sun	Terrestrial Sun	2900°K Tungsten (Flat)	2900°K Tungsten (Coiled, P/D=2)
0.01	0.02	0.02	0.29	0.35	0.44	0.45
0.03	0.02	0.04	0.33	0.39	0.54	0.55
0.05	0.02	0.06	0.36	0.41	0.58	0.59
0.08	0.04	0.10	0.39	0.44	0.65	0.66
0.15	0.10	0.20	0.45	0.48	0.75	0.76
0.25	0.10	0.30	0.52	0.55	0.88	0.90
0.35	0.10	0.40	0.59	0.62	1.01	1.03
0.45	0.10	0.50	0.68	0.68	1.15	1.17
0.55	0.10	0.60	0.79	0.76	1.30	1.32
0.65	0.10	0.70	0.93	0.84	1.48	1.51
0.75	0.10	0.80	1.11	1.04	1.73	1.78
0.85	0.10	0.90	1.43	1.27	2.15	2.21
0.92	0.04	0.94	1.87	1.64	2.71	2.80
0.95	0.02	0.96	2.23	1.84	3.19	3.30
0.97	0.02	0.98	2.70	2.15	3.73	3.89
0.99	0.02	1.00	3.93	3.10	5.34	5.50

$$\sigma T^4 = q_b^+ = \dot{Q}/\pi DL \;, \quad T = \left\{\frac{1850}{(\pi)(.025)(1)(5.6697\text{x}10^{-8})}\right\}^{1/4}$$

$$= 803 \text{ K}$$

Accordingly $\lambda T = (2.7\ \mu m)(803\ K) = 2168\ \mu m\ K$ and $f_e \doteq 0.095$.

Thus the absorbed flux αq^- is

$$\alpha q^- = [(.095)(.30)+(.905)(.95)][2944] = 2615\ W/m^2$$

REFERENCES TO SECTION 1

1. Czerny, M., and A. Walther, Tables of Fractional Functions for the Planck Distribution Law, Springer Verlag, Berlin, 1961.

2. Dunkle, R. V., "Thermal Radiation Tables and Applications", Trans. Am. Soc. Mech. Engrs., Vol. 76, pp. 549-552, 1954.

3. Edwards, D. K., "Radiative Transfer Characteristics of Materials", J. Heat Transfer, Vol. 91, pp. 1-15, 1969.

4. Knuth, E. L., Introduction to Statistical Thermodynamics, McGraw-Hill, New York, 1966 (see Appendix I, "Emissions from Surfaces").

5. Planck, M., The Theory of Heat Radiation, Dover, New York, 1959.

6. Snyder, N. W., "A Review of Thermal-Radiation Constants", Trans. Am. Soc. Mech. Engrs., Vol. 76, pp. 537-539, 1954.

7. Worthing, A. G., Temperature, Its Measurement and Control in Science and Industry, Reinhold, New York, 1941, pp. 1164-1187.

EXERCISES FOR SECTION 1

1. Create a table giving numerical values of the following quantities: (Row 1) total radiosity in W/m^2, (Row 2) spectral intensity at 1000 cm^{-1} in hybrid units of $W/m^2\ cm^{-1}$ (steradian), (Row 3) spectral intensity at 10 μm in hybrid units of W/m^2 μm (steradian), (Row 4) the total radiosity contributed by photons with wavelengths between 10 μm and 11 μm. The preceding quantities are to be for a black-cavity radiator at (Column 1) 300 K and (Column 2) 1000 K.

2. Ordinary window glass a few millimeters thick transmits from approximately 0.33 μm to 2.7 μm. Why or why not is it a reasonable engineering approximation to treat the glass as opaque in analyzing the heat loss from a 300 K room through a window? Can the same be said for the heat loss from a 400 K solar absorber plate through a 300 K coverglass of a solar collector?

3. How many Watts from a 100 W coiled tungsten light bulb filament at 2000 K would be absorbed by a glass envelope opaque to wavelengths longer than 2.7 μm but nonabsorbing to shorter wavelengths?

4. If the sun is approximated as a perfectly diffusely emitting sphere (actually there is what is called "dark-

ening to the limb"), what irradiation would be received on one face of a spacecraft radiator whose plane bisected the sun? Take the irradiation upon a plate whose normal passes through the sun to be 1380 W/m^2 at a distance of one astronomical unit, and take the sun to be a sphere subtending a half angle of 16 minutes of arc.

5. What is the irradiation upon an element of wall parallel to a 50-cm-long 2.5-cm-diameter rod 10 cm away? The rod radiates diffusely 1850 W/m, and the irradiation from the surrounds is negligible. Consider the element to be opposite (a) the center of the rod or (b) one end of the rod.

2 *SURFACE RADIATION CHARACTERISTICS*

2.A Introduction to Surface Characteristics. To the thermal designer, surface radiation characteristics are numbers that must be entered into his computer in order to get answers to the questions he puts to the computer about the size, orientation, or spacing of elements; material selection; and so forth. The designer hopes that values of these characteristics can be found in a handbook or data compilation. At times, however, the handbook leaves the designer unsatisfied. It lists no value for a given material or such a large range of values that the designer realizes that a test measurement is in order. Thereupon it becomes apparent that many different kinds of tests can be conducted; many different numbers can be obtained. Which test or set of tests is to be run? How is a single number to be extracted from the test results? What do the numbers that are listed in the handbook mean? Are they really applicable?

Two categories of radiation characteristics must be distinguished: (1) model characteristics used to model surface behavior in a computer program or calculation scheme and (2) real characteristics that describe how the surface really does behave. Section 2 begins with a review of real characteristics and their measurements, reviews the electromagnetic theory of reflection, and concludes by examining the use of surface characteristics in thermal design.

The nomenclature used for surface radiation characteristics varies considerably in the literature. One question that often occurs is the meaning of an "ivity" ending versus an "ance" ending. Worthing of the U.S. National Bureau of Standards at one time suggested using "ivity" for an intrinsic property of a pure, perfectly-polished and annealed specimen and "ance" for surfaces having roughness, surface oxide layers, or other nonideal features. Various workers have followed the suggestion at one time or another, and others have not. The "ivity" ending used here has no special meaning and may be regarded as interchangeable with the "ance" ending.

2.B <u>Absorption and Emission Characteristics</u>. Briefly put, the absorptivity of a surface system is a property value between zero and unity that defines what fraction of the irradiation is absorbed. The emissivity is a similar property value between zero and unity that defines what fraction of the black body radiosity is actually emitted by a surface system. What are these property value functions of? They clearly depend upon the type of surface system, the material it is made of and its structure as determined by heat treatment, cold work, oxide film thickness, roughness and so forth. If the structure is in a stable configuration (by no means always the case), one regards the properties as functions of its thermodynamic state fixed by its temperature T_s. Moreover the properties depend upon the nature of the thermal radiation in question, specifically its direction and wavelength, and at times its state of polarization is a concern.

Accordingly one distinguishes between directional and directionally-averaged values and spectral and spectrally-averaged values. The word "hemispherical" is reserved for values directionally averaged over the entire 2π steradian hemisphere of solid angle above a surface, and the word "total" is reserved for values spectrally averaged over the total spectrum, i.e., encompassing all wavelengths or wavenumbers from zero to infinity. This distinction is by no means universal; one encounters the term "total" being used in the sense of a totality of directions (i.e. in place of hemispherical) as well as in the sense of a totality of wavelengths.

Let $\alpha(\theta,\phi,\lambda,T_s)$ denote the spectral (λ), directional (θ,ϕ) absorptivity of a surface system at temperature T_s for unpolarized irradiation. Let $\varepsilon(\theta,\phi,\lambda,T_s)$ denote the spectral directional emissivity. Then the net flux leaving surface s is

$$q = \int_0^{\infty} \int_0^{2\pi} \int_0^{\pi/2} \{\varepsilon(\theta,\phi,\lambda,T_s) I_{b\lambda}(\lambda,T_s)$$

$$-\alpha(\theta,\phi,\lambda,T_s) I_\lambda^-(\theta,\phi,\lambda)\} \cos\theta \sin\theta \, d\theta \, d\phi \, d\lambda \tag{2-1}$$

Wavenumber ν and wavelength λ are interchangeable in this and the subsequent relations.

The Principle of Detailed Balancing requires that at thermodynamic equilibrium there be no net energy transfer between two surfaces at the same temperature, not just in aggregate, but in detail for each direction and wavenumber. This principle presents philosophical problems to students just as does the Second Law of Thermodynamics, to which it is related, and these problems are not taken up here. Suffice it to say that when the surrounds are a hohlraum at temperature T_s, then I_λ^- is $I_{b\lambda}(\lambda,T_s)$ in Eq. (2-1), and the Principle of Detailed Balancing requires

$$\varepsilon(\theta,\phi,\lambda,T_s) = \alpha(\theta,\phi,\lambda,T_s) \tag{2-2}$$

This relation is called Kirchhoff's Law.

Even when the environment, say a hohlraum at temperature T_e not equal to T_s, is not in equilibrium with the surface, our assumption that the surface system remains in a thermodynamic equilibrium microstate characterized by temperature T_s is sufficient to keep Eq. (2-2) in force. One may be able to imagine or even contrive to have a nonequilibrium surface state. In such a case Kirchhoff's Law would be inoperative.

Let the environment be a hohlraum at temperature T_e. Then $I_\lambda^-(\theta,\phi,\lambda)$ is $I_{b\lambda}(\lambda,T_e)$, and Eq. (2-1) becomes with Kirchhoff's Law

$$q = \int_0^\infty \int_0^{2\pi} \int_0^{\pi/2} \{\alpha(\theta,\phi,\lambda,T_s) I_{b\lambda}(\lambda,T_s) - \alpha(\theta,\phi,\lambda,T_s) I_{b\lambda}(\lambda,T_e)\} \cos\theta \sin\theta \, d\theta \, d\phi \, d\lambda$$

$$= \varepsilon_{TH}(T_s)\sigma T_s^4 - \alpha_{TH}(T_s,T_e)\sigma T_e^4 \tag{2-3}$$

This expression defines the total hemispherical emissivity and total hemispherical absorptivity respectively. These properties are spectrally and directionally averaged. One can, of course, separate the two averaging steps into a hemispherical averaging one denoted by subscript H

$$\alpha_H(\lambda,T_s) = \frac{1}{\pi} \int_0^{2\pi} \int_0^{\pi/2} \alpha(\theta,\phi,\lambda,T_s) \cos\theta \sin\theta \, d\theta \, d\phi \tag{2-4}$$

and a total spectral averaging denoted by subscript T

$$\varepsilon_{TH}(T_s) = \frac{1}{\sigma T_s^4} \int_0^\infty \alpha_H(\lambda,T_s) \pi I_{b\lambda}(\lambda,T_s) \, d\lambda \tag{2-5}$$

$$\alpha_{TH}(T_s,T_e) = \frac{1}{\sigma T_e^4} \int_0^\infty \alpha_H(\lambda,T_s) \pi I_{b\lambda}(\lambda,T_e) \, d\lambda \tag{2-6}$$

Equation (1-30) shows

$$\frac{\pi I_{b\lambda}(\lambda,T_s)\, d\lambda}{\sigma T_s^4} = df_e(\lambda T_s)$$

Accordingly Eqs. (2-5 and 6) can be written

$$\varepsilon_{TH}(T_s) = \int_0^1 \alpha_H(\lambda,T_s)\,df_e(\lambda T_s) \tag{2-7}$$

$$\alpha_{TH}(T_s,T_e) = \int_0^1 \alpha_H(\lambda,T_s)\,df_e(\lambda T_e) \tag{2-8}$$

Some have made the observation that as T_e approaches T_s, $\alpha_{TH}(T_s,T_e)$ approaches $\varepsilon_{TH}(T_s)$. But this observation is without meaning. When T_e approaches T_s, Eq. (2-3) may be written

$$q = \int_0^\infty \alpha_H(\lambda,T_s)\,[\pi I_{b\lambda}(\lambda,T_s) - \pi I_{b\lambda}(\lambda,T_e)]\,d\lambda$$

$$q = \int_0^\infty \alpha_H(\lambda,T_s)\,[\partial q_b^+(\lambda,T_s)/\partial T_s]\,[T_s - T_e]\,d\lambda$$

$$= \varepsilon_{T_iH}(T_s)\,4\sigma T_s^{\,3}\,[T_s - T_e]$$

where the internal total (and hemispherical) emissivity is

$$\varepsilon_{T_iH}(T_s) \equiv \frac{1}{4\sigma T_s^{\,3}} \int_0^\infty \alpha_H(\lambda,T_s)\,[\partial q_b^+(\lambda,T_s)/\partial T_s]\,d\lambda \tag{2-9}$$

Comparison with Eq. (1-31) shows

$$\varepsilon_{T_iH} = \int_0^1 \alpha_H(\lambda,T_s)\,df_i(\lambda T_s) \tag{2-10}$$

Now the observation can be made that internal-total emissivity $\varepsilon_{T_iH}(T_s)$ and internal-total absorptivity $\alpha_{T_iH}(T_s)$ are indistinguishable.

Note that Kirchhoff's Law as referred to here is understood to be the meaningful Eq. (2-2) and not the vacuous statement that $\varepsilon_{TH}(T_s) = \alpha_{TH}(T_s, T_e{=}T_s)$.

As a concrete example of the use of Eqs. (2-7, 8, and 10) consider that a graph of $\alpha_H(\lambda)$ is available for a given material and a value of T_s. For example, stainless steel at 300 K has a graph of $\alpha_H(\lambda)$ that can be approximated by $\alpha_H(\lambda) \doteq 0.44\lambda^{-1/2}$ for the range 1 μm < λ < 100 μm. Suppose we want the external hemispherical emissivity at 300 K, the internal total hemispherical emissivity at 300 K, and the total hemispherical absorptivity for a 1000 K black body source. Table 2-1 is constructed from Table 1-1 and the given $\alpha_H(\lambda)$ data. The results are $\varepsilon_{TH} = 0.120$, $\varepsilon_{T_iH} = 0.135$, and $\alpha_{TH} = 0.219$.

2.C <u>Measurement of Absorption and Emission Characteristics</u>. As can be seen from the foregoing, one can measure spectral or total and/or directional or hemispherical absorptivity or emissivity, and the total quantities can be internal total or external total values.

When one desires spectral values at high values of T_s one usually measures spectral directional emissivity by placing a heated specimen in a low temperature hohlraum and comparing the intensity I_λ^+ of the specimen with $I_{b\lambda}$ from a reference cavity emitter at close to the same temperature T_s.

TABLE 2-1

SAMPLE CALCULATION OF TOTAL EMISSIVITY AND ABSORPTIVITY

Fractional Function Increment	External Total Emissivity at 300 K				Internal Total Emissivity at 300K				Total Absorptivity for 1000k source			
	λT_s cm K	λ μm	α_H	$\alpha_H \Delta f_e$	λT_s cm K	λ μm	α_H	$\alpha_H \Delta f_i$	λT_e cm K	λ μm	α_H	$\alpha_H \Delta f_e$
0.005	.1230	4.10	.217	.0011	.1073	3.58	.233	.0012	.1230	1.23	.397	.0020
0.005	.1395	4.65	.204	.0010	.1209	4.03	.219	.0011	.1395	1.40	.373	.0019
0.005	.1495	4.98	.197	.0010	.1291	4.30	.212	.0011	.1495	1.50	.360	.0018
0.005	.1573	5.24	.192	.0010	.1352	4.51	.207	.0010	.1573	1.57	.351	.0018
0.01	.1662	5.54	.187	.0019	.1423	4.74	.202	.0020	.1662	1.66	.341	.0034
0.01	.1762	5.87	.182	.0018	.1501	5.00	.197	.0020	.1762	1.76	.332	.0033
0.01	.1848	6.16	.177	.0018	.1565	5.22	.193	.0019	.1848	1.85	.323	.0032
0.01	.1922	6.41	.174	.0017	.1625	5.42	.189	.0019	.1922	1.92	.317	.0032
0.02	.202	6.73	.170	.0034	.1701	5.67	.185	.0037	.202	2.02	.310	.0062
0.02	.214	7.13	.165	.0033	.1791	5.97	.180	.0036	.214	2.14	.301	.0060
0.02	.225	7.50	.161	.0032	.1874	6.25	.176	.0035	.225	2.25	.293	.0059
0.02	.235	7.83	.157	.0031	.1949	6.50	.173	.0035	.235	2.35	.287	.0057
0.02	.245	8.17	.154	.0031	.202	6.73	.170	.0034	.245	2.45	.281	.0056
0.02	.254	8.47	.151	.0030	.209	6.97	.167	.0033	.254	2.54	.276	.0055
0.02	.263	8.77	.149	.0030	.216	7.20	.164	.0033	.263	2.63	.271	.0054
0.02	.272	9.07	.146	.0029	.222	7.40	.162	.0032	.272	2.72	.267	.0053
0.02	.281	9.37	.144	.0029	.229	7.63	.159	.0032	.281	2.81	.262	.0052
0.02	.290	9.67	.142	.0028	.235	7.83	.157	.0031	.290	2.90	.258	.0052
0.02	.299	9.97	.139	.0028	.242	8.07	.155	.0031	.299	2.99	.255	.0051
0.02	.308	10.3	.137	.0027	.248	8.27	.153	.0031	.308	3.08	.251	.0050
0.02	.317	10.6	.135	.0027	.255	8.50	.151	.0030	.317	3.17	.247	.0049
0.02	.326	10.9	.134	.0027	.261	8.70	.149	.0030	.326	3.26	.244	.0049
0.02	.335	11.2	.132	.0026	.268	8.93	.147	.0029	.335	3.35	.240	.0048
0.02	.344	11.5	.130	.0026	.275	9.17	.145	.0029	.344	3.44	.237	.0047
0.02	.353	11.8	.128	.0026	.281	9.37	.144	.0029	.353	3.53	.234	.0047
0.02	.363	12.1	.127	.0025	.288	9.60	.142	.0028	.363	3.63	.231	.0046
0.02	.373	12.4	.125	.0025	.295	9.83	.140	.0028	.373	3.73	.228	.0046
0.02	.384	12.8	.123	.0025	.303	10.1	.138	.0028	.384	3.84	.225	.0045
0.02	.394	13.1	.121	.0024	.310	10.3	.137	.0027	.394	3.94	.222	.0044
0.02	.405	13.5	.120	.0024	.318	10.6	.135	.0027	.405	4.05	.219	.0044
0.02	.416	13.9	.118	.0024	.326	10.9	.134	.0027	.416	4.16	.216	.0043

TABLE 2-1 (Cont.)

Fractional Function Increment	External Total Emissivity at 300 K				Internal Total Emissivity at 300 K				Total Absorptivity for 1000k source			
	λT_s cm K	λ µm	α_H	$\alpha_H \Delta f_e$	λT_s cm K	λ µm	α_H	$\alpha_H \Delta f_i$	λT_e cm K	λ µm	α_H	$\alpha_H \Delta f_e$
0.02	.428	14.3	.117	.0023	.334	11.1	.132	.0026	.428	4.28	.213	.0043
0.02	.440	14.7	.115	.0023	.342	11.4	.130	.0026	.440	4.40	.210	.0042
0.02	.453	15.1	.113	.0023	.351	11.7	.129	.0026	.453	4.53	.207	.0041
0.02	.467	15.7	.112	.0022	.361	12.0	.127	.0025	.467	4.67	.204	.0041
0.02	.482	16.1	.110	.0022	.371	12.4	.125	.0025	.482	4.82	.200	.0040
0.02	.497	16.6	.108	.0022	.382	12.7	.123	.0025	.497	4.97	.197	.0039
0.02	.513	17.1	.106	.0021	.393	13.1	.122	.0024	.513	5.13	.194	.0039
0.02	.531	17.7	.105	.0021	.405	13.5	.120	.0024	.531	5.31	.191	.0038
0.02	.550	18.3	.103	.0021	.417	13.9	.118	.0024	.550	5.50	.188	.0038
0.02	.570	19.0	.101	.0020	.430	14.3	.116	.0023	.570	5.70	.184	.0037
0.02	.593	19.8	.099	.0020	.446	14.9	.114	.0023	.593	5.93	.181	.0036
0.02	.616	20.5	.097	.0019	.462	15.4	.112	.0022	.616	6.16	.177	.0035
0.02	.642	21.4	.095	.0019	.479	16.0	.110	.0022	.642	6.42	.174	.0035
0.02	.671	22.4	.093	.0019	.498	16.6	.108	.0022	.671	6.71	.170	.0034
0.02	.704	23.5	.091	.0018	.521	17.4	.106	.0021	.704	7.04	.166	.0033
0.02	.741	24.7	.089	.0018	.546	18.2	.103	.0021	.741	7.41	.162	.0032
0.02	.783	26.1	.086	.0017	.574	19.1	.101	.0020	.783	7.83	.157	.0031
0.02	.825	27.5	.084	.0017	.609	20.3	.098	.0020	.825	8.25	.153	.0031
0.02	.899	30.0	.080	.0016	.653	21.8	.094	.0019	.899	8.99	.147	.0029
0.02	.982	32.7	.077	.0015	.690	23.0	.092	.0018	.982	9.82	.140	.0028
0.02	1.090	36.3	.073	.0015	.774	25.8	.087	.0017	1.090	10.9	.133	.0027
0.01	1.200	40.0	.070	.0007	.846	28.2	.083	.0008	1.200	12.0	.127	.0013
0.01	1.298	43.3	.067	.0007	.909	30.3	.080	.0008	1.298	13.0	.122	.0012
0.01	1.433	47.8	.064	.0006	.997	33.2	.076	.0008	1.433	14.3	.116	.0012
0.01	1.63	54.3	.060	.0006	1.123	37.4	.072	.0007	1.63	16.3	.109	.0011
0.005	1.87	62.3	.056	.0003	1.28	42.7	.067	.0003	1.87	18.7	.102	.0005
0.005	2.22	74.0	.051	.0003	1.43	47.7	.064	.0003	2.22	22.2	.093	.0005
0.005	2.56	85.3	.048	.0002	1.70	56.7	.059	.0003	2.56	25.6	.087	.0004
0.005	7.34	245.0	.028	.0001	2.49	83.0	.048	.0002	7.34	73.4	.051	.0003
				.120				.135				.219

When one desires spectral values at moderate or low values of T_s one usually makes a determination of spectral directional reflectivity $\rho(\theta,\phi,\lambda,T_s)$ (as discussed in Section 2.D) and by subtraction obtains (for an opaque specimen)

$$\varepsilon(\theta,\phi,\lambda,T) = \alpha(\theta,\phi,\lambda,T) = 1-\rho(\theta,\phi,\lambda,T) \qquad (2\text{-}11)$$

For a transparent specimen, reflectivity and transmissivity can be determined and absorptivity found by subtraction from unity.

For spectral measurements a filter, prism, grating spectrometer, or interferometer is used. The instrument usually views the specimen within a narrow solid angle about θ and ϕ to yield directional quantities, but it is possible to use a hemispherical or masked cylindrical specimen to obtain a hemispherical measurement directly.

For total emissivity measurements one can use a total radiometer having a detector that responds to the total radiant heating effect upon it. One can view a flat or hemispherical specimen, or one can make a calorimetric determination by heating a specimen suspended within a low temperature evacuated hohlraum. Calorimetric measurements of absorptivity can be similarly made by irradiating such a specimen with high temperature black body radiation.

One should note that a laboratory emissivity determination corresponds exactly to neither the (external) total emissivity defined by Eq. (2-5 or 7) nor to the internal total emissivity defined by Eq. (2-9 or 10) when the temperature of the specimen environment T_e is neither far removed from nor very close to the specimen temperature T_s. The laboratory observation is usually one of

$$\alpha_{T_oH}(T_s,T_e) = \frac{\int_0^\infty \alpha_H(\lambda,T_s)\,[\pi I_{b\lambda}(\lambda,T_s)-\pi I_{b\lambda}(\lambda,T_e)]\,d\lambda}{\sigma T_s^4-\sigma T_e^4} \tag{2-12}$$

Only when T_e is very small compared to T_s does Eq. (2-12) approximate Eq. (2-5); only when T_e is close to T_s does it approximate Eq. (2-9). For this reason, and for other reasons to be made clear in subsequent sections, spectral data are desirable in addition to (or even instead of) total observations.

2.D <u>Reflection and Transmission Characteristics</u>. One often wants to know what fraction of the irradiation q^- becomes a contribution to the radiosity q^+ on one side or the other of a wall. That fraction pertaining to the irradiated side is termed reflectivity, and that pertaining to the other side is the transmissivity. Clearly these fractions will be properties of the wall material and structure and its thermodynamic state and also the spectral and directional distribution of the irradiation. The terms spectral and total apply to the reflectivity and transmissivity in the same sense as to the absorptivity. The terms directional and hemispherical also apply as before. Accordingly the spectral directional reflectivity is defined as the reflected portion of the surface radiosity q_R^+ as a fraction of the spectral irradiation $I^- \cos^- \Delta\Omega^-$

$$\rho_\lambda(\theta^-, \phi^-, \lambda, T_s) = \frac{q_{R\lambda}^+}{I_\lambda^-(\theta^-, \phi^-, \lambda) \cos\theta^- \Delta\Omega^-} \tag{2-13}$$

Spectral directional transmissivity $\tau(\theta^-, \phi^-, \lambda, T_s)$ is similarly defined, with an additional caveat that the entire wall be near a state of thermodynamic equilibrium characterized by temperature T_s. The idea of an m-surface as shown in Fig. 1-1 must be abandoned for a transmitting wall. No provision has been made in these definitions for phosphorescence or fluorescence.

Note that these directional characteristics carry no information about the directional distribution of the reflected and transmitted portions of the irradiation. If one wants to know not just the reflected portion of the radiosity q_R^+ but its intensity I_R^+, one requires a characteristic that varies not just with direction of incidence θ^-, ϕ^-

but also with direction of emergence θ^+,ϕ^+. For a perfectly diffuse surface $q^+_{R\lambda}$ is $\pi I^+_{R\lambda}$, of course. Accordingly a spectral bidirectional reflectivity may be defined as

$$\rho_{B\lambda}(\theta^+,\phi^+,\theta^-,\phi^-,\lambda,T_s) = \frac{\pi I^+_R\ (\theta^+,\phi^+,\lambda)}{I^-(\theta^-,\phi^-,\lambda)\cos\theta^-\Delta\Omega^-} \tag{2-14}$$

This quantity finds its main application in thermal engineering in the design of instruments for property determination and in the understanding of the operation of such instruments. It is also of use in problems of thermometry and remote sensing via heat radiation.

If one considers the radiant transfer from say a black surface at temperature T_1 to another black one at temperature T_2 via reflection from a nonblack surface at T_s and the transfer from 2 to 1 via the same route, and if one then lets T_1 approach T_2 and invokes the Principle of Detailed Balancing, one is led to the Helmholtz Reciprocity Principle

$$\rho_{B\lambda}(\theta_1,\phi_1,\theta_2,\phi_2,\lambda,T_s) = \rho_{B\lambda}(\theta_2,\phi_2,\theta_1,\phi_1,\lambda,T_s) \tag{2-15}$$

The Reciprocity Principle has important consequences in instrument design and application and is a constraint upon builders of mathematical models for bidirectional reflection behavior of surfaces.

One understands by perfectly diffuse reflection the situation where (as a limiting ideal) the bidirectional reflectivity is independent of direction. For a perfectly diffuse surface there is no distinction between hemispherical, directional, or bidirectional reflectivity. The same numerical value applies to all three characteristics. One understands by imperfectly diffuse reflection the situation where the bidirectional reflectivity defined according to Eq. (2-14) is finite even in the limit as $\Delta\Omega^-$ goes to zero. In such a case it is clear from Eqs. (2-13 and 14) and Eq. (1-8) that the directional reflectivity is

$$\rho_\lambda(\theta^-,\phi^-,\lambda,T_s) = \frac{1}{\pi}\int_0^{2\pi}\int_0^{\pi/2} \rho_\lambda(\theta^+,\phi^+,\theta^-,\phi^-,\lambda,T_s) \times \cos\theta^+ \sin\theta^+ d\theta^+ d\phi^+ \tag{2-16}$$

and the hemispherical reflectivity is

$$\rho_{H\lambda}(\lambda,T_s) = \frac{1}{\pi}\int_0^{2\pi}\int_0^{\pi/2} \rho_\lambda(\theta^-,\phi^-,\lambda,T_s) \times \cos\theta^- \sin\theta^- d\theta^- d\phi^- \tag{2-17}$$

$$\rho_{H\lambda}(\lambda,T_s) = \frac{1}{\pi}\int_0^{2\pi}\int_0^{\pi/2}\frac{1}{\pi}\int_0^{2\pi}\int_0^{\pi/2} \rho_\lambda(\theta^+,\phi^+,\theta^-,\phi^-,\lambda,T_s)\cos\theta^- \sin\theta^- d\theta^- d\phi^- \times \cos\theta^+ \sin\theta^+ d\theta^+ d\phi^+ \tag{2-18}$$

Specular reflection gives rise to a reflected intensity that lies entirely within a solid angle $\Delta\Omega^+$ about emergent direction θ^+,ϕ^+ equal to the solid angle $\Delta\Omega^-$ about incident direction θ^-,ϕ^- where $\theta^+ = \theta^-$, and $\phi^+ = \phi^- + \pi$ (measuring both ϕ^+ and ϕ^- from the same reference x-axis as shown in Fig. 1-2). The reflected intensity I_R^+ is a fraction of the incident I^- fixed by the specular reflectivity

$$\rho_S(\theta^-,\phi^-,\lambda,T_s) = \frac{I_{R\lambda}^+}{I_\lambda^-} \tag{2-19}$$

This definition corresponds exactly to Eq. (2-13) because of the equality in $\cos\theta^+\Delta\Omega^+$ and $\cos\theta^-\Delta\Omega^-$. Equation (2-17) for the hemispherical reflectivity applies as well.

2.E <u>Measurement of Reflection and Transmission Characteristics</u>. A few of the considerations involved in making such measurements are described briefly to give the thermal designer some appreciation for the types of measurements usually employed.

One common consideration is the ability to distinguish reflected or transmitted radiation from emitted radiation. It may be that $I_b(\lambda,T_s)$ is so low by reason of a low temperature T_s that it is negligible compared to the incident or reflected intensities. If it is not, it may be possible to tag the incident radiation by chopping. The irradiation $I^-\cos\theta^-\Delta\Omega^-$ is interrupted or modulated in a steady periodic manner, and the signal from the detector is processed to record only the radiation in phase with the irradiation. In this way the thermal emission, which is steady by virtue of the thermal capacity of the specimen, is rejected. The phasing of the source chopper and detection equipment must of course be correct so that the signal recorded is independent of sample temperature, as may be verified by increasing sample temperature with the irradiation shuttered off and observing no change in the zero signal.

In making measurements of directional reflectivity and/or transmittivity one finds that it is often much easier to provide uniform irradiation (I^- independent of θ^-,ϕ^-) than to achieve uniform detection ($I^+ \cos\theta^+ d\Omega^+$ detected equally well independent of direction). One then makes use of the Helmholtz Reciprocity Principle to interchange the plus- and minus-superscripted angles in Eq. (2-16) with the result that $I^+_{R\lambda}(\theta^+,\phi^+,\lambda)$ is observed and compared to the uniform I^-_λ. The observed quantity (which may be called the hemispherical-directional reflectivity following Birkebak and Eckert [27]) is identical to the directional (or hemispherical-directional) quantity defined by Eq. (2-13). This fact is the basis for the heated cavity reflectometer [12-15], the reciprocal-mode 2π-steradian mirror reflectometer [18-23], and the reciprocal-mode integrating sphere [8-10].

For example, in the heated cavity reflectometer [12,15] a water- or gas-cooled specimen is exposed to a 1000 K cavity. The irradiation consequently has a uniform intensity $I_{b\lambda}(T_c)$, where T_c is the cavity temperature, except for the small solid angle subtended by the view port. See Fig. 2-1. The small view port permits the intensity from the specimen to be viewed within a small solid angle $\Delta\Omega$ about direction θ,ϕ. The signal V_s when the detector views the sample less the signal V_o when the detector views a cavity at the sample temperature is determined. It is done by having a chopper with a highly reflecting specular blade and directing the reflected line of sight from the detector into a cavity at the sample temperature T_s. When the sample is replaced by a platinum fin at the main cavity temperature T_c the intensity difference $I_{b\lambda}(T_c)-I_{b\lambda}(T_s)$ gives rise to signal V_r-V_o. Accordingly

$$\frac{V_s-V_o}{V_r-V_o} = \frac{I^+_{s\lambda}-I_{b\lambda}(T_s)}{I_{b\lambda}(T_c)-I_{b\lambda}(T_s)}.$$

But

$$I^+_{s\lambda} = \varepsilon_\lambda(\theta,\phi,\lambda,T_s)I_{b\lambda}(T_s)+\rho_\lambda(\theta,\phi,\lambda,T_s)I_{b\lambda}(T_c)$$

$$I^+_{s\lambda} = [1-\rho_\lambda(\theta,\phi,\lambda,T_s)]I_{b\lambda}(T_s)+\rho_\lambda(\theta,\phi,\lambda,T_s)I_{b\lambda}(T_c)$$

The result is

$$\frac{V_s-V_o}{V_r-V_o} = \rho_\lambda(\theta,\phi,\lambda,T_s)\,\frac{I_{b\lambda}(T_c)-I_{b\lambda}(T_s)}{I_{b\lambda}(T_c)-I_{b\lambda}(T_s)}$$

$$\frac{V_s-V_o}{V_r-V_o} = \rho_\lambda(\theta,\phi,\lambda,T_s) \qquad (2\text{-}20)$$

When a perfectly specular specimen is oriented with $\theta = 0°$ the exit port is viewed. If this port leads to an enclosure at temperature T_s a signal of zero is received for $V_s - V_o$. When a perfectly diffuse specimen is turned toward the exit port by setting $\theta = 0$, no difference is detected. The exit port view factor is changed by the change in $\cos\theta$ which is nearly unity whether θ is zero or a small angle from zero. Thus the small change in the irradiation on a diffuse specimen is not detected.

In the reciprocal mode 2π-steradian-mirror reflectometer [18-23] the specimen is uniformly irradiated from a cavity (or any continuous-spectrum source) via a mirror. In the reciprocal-mode integrating sphere [8-10] the diffuse irradiation is provided by a highly reflecting diffuse sphere wall which is uniformly irradiated. Interreflections in the sphere serve to smooth out irregularities in the initial irradiation on the sphere wall. In these instruments source-side chopping is readily achieved as shown in Fig. 2-2.

2.F <u>Electromagnetic Theory and the Fresnel Relations.</u> In the classical limit a stream of photons forms a continuous electromagnetic wave with electric field strength $\underline{E}$ and magnetic field strength $\underline{H}$ (the underscore denotes a vector quantity). Maxwell's equations and the so-called constitutive relations for an isotropic medium govern these quantities. Electromagnetic theory offers a description of

FIGURE 2-1

Heated Cavity Reflectometer

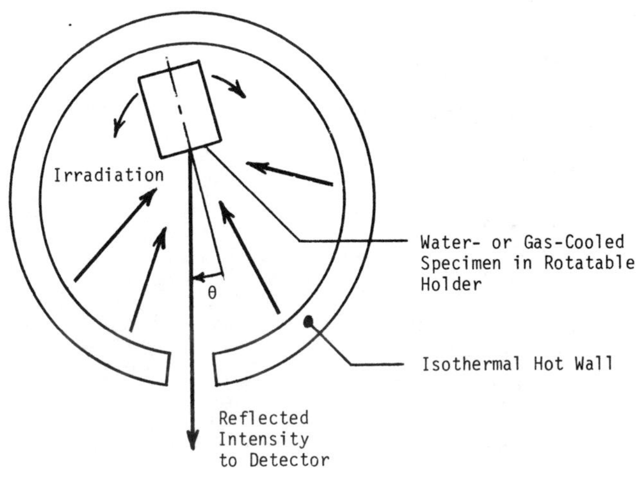

FIGURE 2-2

Reciprocal-Mode Integrating Sphere

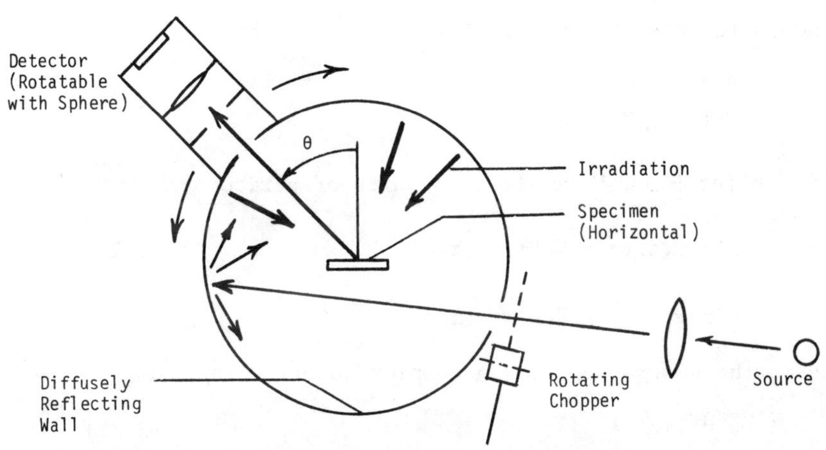

NOTE: The source and entrance port are shown rotated 90° from the end and displaced downwards for clarity.

the specular reflection from a smooth interface or for a number of such interfaces forming surfacing films.

The Maxwell equations and constitutive relations [34] are

$$\nabla \times \underline{E} = -\partial \underline{B}/\partial t \tag{2-21}$$

$$\nabla \times \underline{H} = \underline{j} + \partial \underline{D}/\partial t \tag{2-22}$$

$$\nabla \cdot D = \rho_e \tag{2-23}$$

$$\nabla \cdot B = 0 \tag{2-24}$$

$$\underline{j} = \sigma_e \underline{E} \tag{2-25}$$

$$\underline{D} = \varepsilon_e \underline{E} \tag{2-26}$$

$$\underline{B} = \mu_m \underline{H} \tag{2-27}$$

where $\underline{B}$ is the magnetic induction, $\underline{j}$ the electric current density, $\underline{D}$ the electric induction, ρ_e the electric charge density, σ_e the electrical conductivity, ε_e the electric permittivity, and μ_m the magnetic permeability. The velocity of light in vacuum (denoted by subscript $_0$) is

$$c_0^{\,2} = \frac{1}{\varepsilon_0 \mu_0} \tag{2-28}$$

Let $\underline{E}$ and $\underline{H}$ be the real parts of periodic variations

$$\underline{E}(x,y,z,t) = \mathcal{R}\{\tilde{\underline{E}}(x,y,z)e^{i\omega t}\} \tag{2-29}$$

$$\underline{H}(x,y,z,t) = \mathcal{R}\{\tilde{\underline{H}}(x,y,z)e^{i\omega t}\} \tag{2-30}$$

where the super $\tilde{}$ denotes a complex quantity, and

$$\omega = 2\pi c_0/\lambda_0 \tag{2-31}$$

Electric neutrality requires

$$\rho_e = 0 \tag{2-32}$$

Accordingly Eqs. (2-21 to 27) give a time independent set of equations

$$\nabla x \underline{\tilde{E}} = -i\omega\mu_m \underline{\tilde{H}} \tag{2-33}$$

$$\nabla x \underline{\tilde{H}} = (\sigma_e + i\omega\varepsilon_e)\underline{\tilde{E}} \tag{2-34}$$

$$\nabla \cdot \underline{\tilde{E}} = 0 \tag{2-35}$$

$$\nabla \cdot \underline{\tilde{H}} = 0 \tag{2-36}$$

This set of equations is reduced by taking the curl of Eq. (2-33) and using Eq. (2-34) to eliminate $\underline{\tilde{H}}$.

$$\nabla x \nabla x \underline{\tilde{E}} = -i\omega\mu_m(\sigma_e + i\omega\varepsilon_e)\underline{\tilde{E}}$$

$$\nabla x \nabla x \underline{\tilde{E}} = \omega^2\mu_m\tilde{\varepsilon}_e\underline{\tilde{E}} \tag{2-37}$$

where

$$\tilde{\varepsilon}_e = \varepsilon_e - i\sigma_e/\omega \tag{2-38}$$

By virtue of the identity

$$\nabla x \nabla x \underline{\tilde{E}} = \nabla(\nabla \cdot \underline{\tilde{E}}) - \nabla^2\underline{\tilde{E}}$$

and with Eq. (2-15) one obtains

$$\nabla^2\underline{\tilde{E}} = -\omega^2\mu_m\tilde{\varepsilon}_e\underline{\tilde{E}} \tag{2-39}$$

Let the lengths be made dimensionless by ω/c_o, and denote a dimensionless quantity with subscript d. Then Eq. (2-39) becomes

$$\nabla_d{}^2\underline{\tilde{E}} = -\tilde{n}^2\underline{\tilde{E}} \tag{2-40}$$

where $\tilde{n}$ is the complex index of refraction

$$\tilde{n}^2 = (\mu_m/\mu_o)(\tilde{\varepsilon}_e/\varepsilon_o) = (n-ik)^2 \tag{2-41}$$

The quantity n is called the refractive index and the quantity k the absorptive index. Collectively they are referred to as the optical constants of the medium. They are in general dependent upon wavelength or frequency of the radiation. Note that in vacuum n = 1 and k = 0. A material with a zero value of k is one with a zero value of σ_e, hence a dielectric.

The governing equation for $\underline{\tilde{E}}$ admits a solution periodic in space. Consider a wave in the first quadrant of the x-z plane

$$\partial\tilde{E}/\partial x_d = -i\tilde{a}\underline{\tilde{E}} \ , \quad \partial\tilde{E}/\partial y_d = 0 \ , \quad \partial\tilde{E}/\partial z_d = -i\tilde{b}\tilde{E}$$

$$\underline{\tilde{E}} = \underline{\tilde{A}}e^{i(\omega t-\tilde{a}x_d-\tilde{b}z_d)} \tag{2-42}$$

Equation (2-40) requires that

$$\tilde{a}^2 + \tilde{b}^2 = \tilde{n}^2 \tag{2-43}$$

Accordingly, parameter $\tilde{\theta}$ is introduced where

$$\tilde{a} = \tilde{n}\sin\tilde{\theta} \qquad \tilde{b} = \tilde{n}\cos\tilde{\theta} \tag{2-44}$$

The result is that the governing electromagnetic equations admit a solution in the form of a wave

$$\underline{E} = \mathcal{R}\{\underline{\tilde{A}}e^{i(\omega t-\tilde{n}\sin\tilde{\theta}x_d-\tilde{n}\cos\tilde{\theta}z_d)}\} \tag{2-45}$$

The magnetic field $\underline{H}$ follows from Eqs. (2-30 and 34).

To tie this development into our notion of radiant intensity and flux, the power per unit area in direction of unit vector $\underline{r}$ is

$$Id\Omega\underline{r} = \frac{1}{4}[\underline{\tilde{E}}x\underline{\tilde{H}}^* + \underline{\tilde{E}}^*x\underline{\tilde{H}}] = \frac{1}{2}\mathcal{R}\{\underline{\tilde{E}}x\underline{\tilde{H}}^*\} \tag{2-46}$$

where the super * denotes a complex conjugate, i.e., if $\tilde{z} = x+iy$, $\tilde{z}^* = x-iy$.

To determine the reflectivity of an interface it is necessary to know the boundary conditions required to be satisfied at an interface. They are continuity of tangential components, continuity of magnetic flux, and conservation of electric charge. Let the interface have normal $\underline{N}$, and let subscripts one and two denote either side

$$\underline{N}x(\tilde{\underline{E}}_2-\tilde{\underline{E}}_1) = 0 \tag{2-47}$$

$$\underline{N}x(\tilde{\underline{H}}_2-\tilde{\underline{H}}_1) = 0 \tag{2-48}$$

$$\underline{N}\cdot(\tilde{\underline{B}}_2-\tilde{\underline{B}}_1) = 0 \tag{2-49}$$

$$\underline{N}\cdot(\tilde{\underline{j}}_2-\tilde{\underline{j}}_1) = -\frac{\partial}{\partial t}(N\cdot(\tilde{\underline{D}}_2-\tilde{\underline{D}}_1)) \tag{2-50}$$

Now consider an approaching wave with amplitude A (see Fig. 2-3) traveling in vacuum (medium 0) and encountering an interface over medium 1. The vector $\tilde{\underline{A}}$ is resolved into components parallel to the plane of incidence (defined by the approaching ray $\underline{r}$ and surface normal $\underline{N}$) and perpendicular to the plane of incidence. Subscripts p and s (from the German parallel and senkrecht) are used. A reflected ray with amplitude $\tilde{\underline{R}}$ and a transmitted one with amplitude $\tilde{\underline{T}}$ are imagined to exist with parameters θ_0' and θ_1 respectively. The six requirements represented by Eqs. (2-47 through 50) lead to six required conditions. Two are referred to as Snell's Law

$$n_0\sin\theta_0 = n_0\sin\theta_0' = \tilde{n}_1\sin\tilde{\theta}_1 \tag{2-51}$$

$$\cos\theta_0 = -\cos\theta_0' \tag{2-52}$$

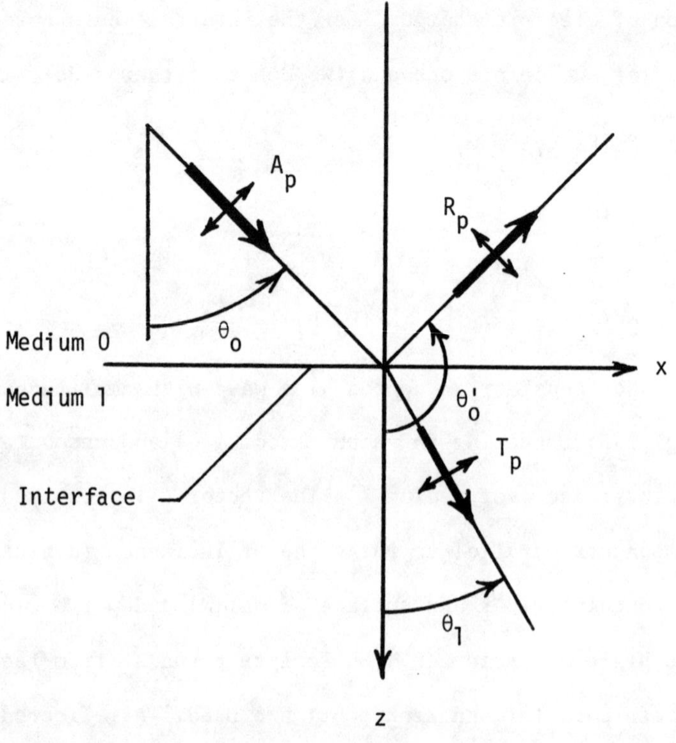

FIGURE 2-3

Approaching, Transmitted, and Reflected Waves at an Interface

The other four lead to the Fresnel relations (subject to the additional assumption that $\mu_1 = \mu_o$).

$$\tilde{r}_{p,0-1} = \frac{\tilde{R}_p}{\tilde{A}_p} = \frac{\tilde{n}_1 \cos\theta_0 - n_0 \cos\tilde{\theta}_1}{\tilde{n}_1 \cos\theta_0 + n_0 \cos\tilde{\theta}_1} \tag{2-53}$$

$$\tilde{t}_{p,0-1} = \frac{\tilde{T}_p}{\tilde{A}_p} = \frac{2n_0 \cos\theta_0}{\tilde{n}_1 \cos\theta_0 + n_0 \cos\tilde{\theta}_1} \tag{2-54}$$

$$\tilde{r}_{s,0-1} = \frac{\tilde{R}_s}{\tilde{A}_s} = \frac{n_0 \cos\theta_0 - \tilde{n}_1 \cos\tilde{\theta}_1}{n_0 \cos\theta_0 + \tilde{n}_1 \cos\tilde{\theta}_1} \tag{2-55}$$

$$\tilde{t}_{s,0-1} = \frac{\tilde{T}_s}{\tilde{A}_s} = \frac{2n_0 \cos\theta_0}{n_0 \cos\theta_0 + \tilde{n}_1 \cos\tilde{\theta}_1} \tag{2-56}$$

If a subsequent interface is encountered, another set of relations is obtained by replacing 0 with 1 and 1 with 2 in the subscripts. The quantity n_0 becomes $\tilde{n}_1$, and $\cos\theta_0$ becomes $\cos\tilde{\theta}_1$, in this replacement.

From Eq. (2-46) one determines that the irradiation in the approaching wave is

$$I^- d\Omega \underline{r} = (I_p^- + I_s^-) d\Omega \underline{r} \tag{2-57}$$

$$I_p^- \cos\theta_0 \, d\Omega = \frac{n_0}{\mu_0 c_0} \cos\theta_0 A_p^2 \tag{2-58}$$

$$I_s^- \cos\theta_0 \, d\Omega = \frac{n_0}{\mu_0 c_0} \cos\theta_0 A_s^2 \tag{2-59}$$

The power per unit area in the reflected wave is I_R^+ where similarly

$$I_R^+ = I_{R,p}^+ + I_{R,s}^+ \tag{2-60}$$

Reflectivities ρ_p and ρ_s may be defined as follows:

$$\rho_p = \frac{I^+_{R,p}}{I^-_p} = \frac{\frac{n_0}{\mu_0 c_0} |\cos\theta_0'| \tilde{R}_p \tilde{R}^*_p}{\frac{n_0}{\mu_0 c_0} \cos\theta_0 A_p^2} = \tilde{r}_p \tilde{r}^*_p \tag{2-61a}$$

$$\rho_s = \frac{I^+_{R,s}}{I^-_s} = \tilde{r}_s \tilde{r}^*_s \tag{2-61b}$$

where the subscripts 0-1 have been omitted for simplicity in notation. Similarly the transmitted to incident power may be found.

$$\tau_p = \mathcal{R}\{\frac{\tilde{n}^* \cos\tilde{\theta} \tilde{t}_p \tilde{t}^*_p}{n_0 \cos\theta_0}\} \tag{2-62a}$$

$$\tau_s = \mathcal{R}\{\frac{\tilde{n}^* \cos\tilde{\theta}^* \tilde{t}_s \tilde{t}^*_s}{n_0 \cos\theta_0}\} \tag{2-62b}$$

Note the difference in the $\cos\tilde{\theta}$ and $\cos\tilde{\theta}^*$ terms.

Since the interface itself is infinitesimally thick, it has no absorptivity, and $\tau_p = 1-\rho_p$ and $\tau_s = 1-\rho_s$. However, if the medium 1 is absorbing and infinite in extent, the transmitted radiation is eventually absorbed within the medium; τ_p and τ_s of the interface become α_p and α_s of the surface system.

Reference [2] develops the Fresnel coefficients $\tilde{r}_{0-2}$ etc. for a surface film made of two interfaces, the 0 -1 and 1-2 interfaces a distance δ_1 apart. By repeated application of the formula $\tilde{r}_{0-n}$ may be constructed by an embedding technique. One begins by obtaining the 0-1 and

1-2 Fresnel coefficients and combines them to obtain the 0-2. Then the 0-2 values replace 0-1, and 2-3 values replace 1-2 values in the formula to yield 0-3 quantities and so on. The formulas for $\tilde{r}_{p,0-2}$ and $\tilde{t}_{p,0-2}$ are

$$\tilde{r}_{p,0-2} = \tilde{r}_{p,0-1} + \frac{\tilde{m}_1^{\,2}\tilde{t}_{p,0-1}\tilde{r}_{p,1-2}\tilde{t}_{p,1-0}}{1-\tilde{m}_1^{\,2}\tilde{r}_{p,1-2}\tilde{r}_{p,1-0}} \tag{2-63a}$$

$$\tilde{t}_{p,0-2} = \frac{\tilde{t}_{p,0-1}\tilde{m}_1\tilde{t}_{p,1-2}}{1-\tilde{m}_1^{\,2}\tilde{r}_{p,1-2}\tilde{r}_{p,1-0}} \tag{2-63b}$$

where

$$\tilde{m}_1^{\,2} = e^{-i4\pi\tilde{n}_1\cos\tilde{\theta}_1\delta_1/\lambda_0} \tag{2-64}$$

The relations for the s-polarized radiation are obtained by replacing p with s in the subscripts above. Reflectivities ρ_p and ρ_s for the 0-2 system are obtained by putting Eq. (2-63a) and the corresponding s-relation into Eqs. (2-61a and b).

An important effect is interference. When δ_1/λ_0 is one value, the reflectivity ρ_p will be a maximum. Excellent mirrors can be made by building up a stack of thin films with appropriate optical properties n and k and thicknesses δ. When δ/λ_0 is another value the reflectivity will be a minimum. High absorptivity or emissivity surfaces can be made in this way. Absorbing surfaces for solar heat collectors and antireflection coatings for glass are so made.

When the surface film is very thick compared to the wavelength, minor variations in thickness and in wavelength result in high and low values of reflectivity due to the interference phenomenon. The engineer should use the appropriate average value obtained by averaging ρ_p and ρ_s over wavelength or thickness, as shown in Ref. [2].

2.G Some Useful Approximations. A number of useful relations can be derived from the foregoing relations when values of optical constants are approximated. For some dielectrics and some semiconductors the refractive index n may be reasonably approximated as a constant over an appreciable portion of the spectrum, and the absorptive index k is small. Furthermore, for normal incidence $\cos\theta_0$ is 1 and $\sin\theta_0$ is zero. Equations (2-61a and b) together with Eqs. (2-53 and 55) reduce to merely

$$\rho_p(\theta=0) = \rho_s(\theta=0) = \left(\frac{n-1}{n+1}\right)^2 \qquad (2\text{-}65)$$

While k is small, the thickness of the surface system must be sufficiently great to make it opaque. In this case the absorptivity and hence the emissivity is

$$\alpha(\theta=0) = 1-\rho(\theta=0) = \frac{4n}{(n+1)^2} \qquad (2\text{-}66)$$

This relation shows that the absorptivity is unity when n = 1 and falls with increasing n.

For unpolarized incident radiation (to be explained more fully in Sec. 2.H), $I_p = I_s = I/2$, and

$$\rho(\theta) = [1/2][\rho_p(\theta)+\rho_s(\theta)] , \tag{2-67}$$

$$\alpha(\theta) = [1/2][\alpha_p(\theta)+\alpha_s(\theta)] \tag{2-68}$$

Eq. (2-4) may be used to obtain the hemispherical absorptance from the directional values fixed by Eqs. (2-68), (2-61a and b), (2-53 and 55), and (2-51). Figure 2-4 from Dunkle [36] shows a plot of $\alpha_H/\alpha(\theta=0)$ for a dielectric.

The figure also shows the ratio for electrical conductors with parameter k/n. Here the approximation is made that $\cos\tilde{\theta}_1 \doteq 1$, because n and k are large in Eq. (2-51). The equations reduce to

$$\rho_p(\theta) \doteq \frac{(n\cos\theta_0-1)^2 + k^2\cos^2\theta_0}{(n\cos\theta_0+1)^2 + k^2\cos^2\theta_0} \tag{2-69}$$

$$\rho_s(\theta) \doteq \frac{(\cos\theta_0-n)^2 + k^2}{(\cos\theta_0+n)^2 + k^2} \tag{2-70}$$

For normal incidence

$$\rho(\theta=0) = \frac{(n-1)^2 + k^2}{(n+1)^2 + k^2} \tag{2-71}$$

$$\alpha(\theta=0) = \frac{4n}{(n+1)^2 + k^2} \tag{2-72}$$

It is clear how these relations may be used together with Eq. (2-4) to find the curves in the figure.

The Hagen-Rubens relation is a crude approximation suitable for some metal alloys. It relates electrical resistivity r_e to optical constants n and k [35].

$$n = k = [30\lambda/r_e]^{1/2} \tag{2-73}$$

When this relation is substituted into Eq. (2-72) and $n >> 1$, one obtains

$$\alpha_\lambda(\theta=0,\lambda,T_s) = 2/n = [2r_e/15\lambda]^{1/2} \tag{2-74}$$

The temperature dependency comes about through its effect upon r_e. Often r_e increases directly with the absolute temperature with the result that

$$\alpha_\lambda(\theta=0,\lambda,T_s) = C[T_s/\lambda]^{1/2} \tag{2-75}$$

where

$$C = [2r_e(T_o)/15T_o]^{1/2} \tag{2-76}$$

If one now finds the total absorptivity for environmental temperature T_e, one finds

$$\alpha_T(\theta=0,T_s,T_e) = CT_s^{1/2}\int_0^1 \lambda^{-1/2}df_e(\lambda T_e)$$

$$\alpha_T(\theta=0,T_s,T_e) = CT_s^{1/2}T_e^{1/2}\int_0^1 u^{-1/2}df_e(u) \tag{2-77}$$

where $f_e(u)$ is the external fractional function. The definite integral is simply a constant (times $\sqrt{k/hc}$)

$$A_e = \int_0^1 u^{-1/2}df_e(u) = 1/(\lambda_e T_e)^{1/2} \tag{2-78}$$

where

$$f_e(\lambda_e T_e) \doteq 0.49\ , \quad \lambda_e T_e \doteq 4050\ \mu\text{m K} \tag{2-79}$$

Therefore

$$\alpha_T(\theta=0,T_e,T_s) = A_e C(T_s T_e)^{1/2} \quad (2\text{-}80)$$

$$\varepsilon_T(\theta=0,T_e) = A_e C T_s \quad (2\text{-}81)$$

$$\varepsilon_{T_i}(\theta=0,T_s) = A_i C T_s \quad (2\text{-}82)$$

where

$$A_i = \int_0^1 u^{-1/2} df_i(u) = 1/(\lambda_i T_s)^{1/2} \quad (2\text{-}83)$$

$$f_i(\lambda_i T_s) \doteq 0.49 \;, \quad \lambda_i T_s \doteq 3180 \ \mu\text{m K} \quad (2\text{-}84)$$

These results can be restated in words to emphasize their meaning. The total emissivity of a bare (i.e. not heavily oxidized) metal alloy increases directly with absolute temperature. The internal total emissivity of a bare metal alloy is approximately one-eighth greater than its external total emissivity. The total absorptivity of a bare metal alloy can be found from its total emissivity by taking the emissivity at the environmental temperature T_e and multiplying by $(T_s/T_e)^{1/2}$. The total absorptivity or emissivity can be estimated from a spectral curve by taking the spectral value of a wavelength where the appropriate fractional function is 49 per cent.

All these results apply to the normal ($\theta_o = 0$) total emissivity and absorptivity, but Fig. 2-4 can be used to correct approximately to hemispherical values using $k/n = 1$ (the curve applies strictly to the spectral values and only approximately to total values).

As has been said, the Hagen-Rubens formula works well only for high resistivity metals, usually metal alloys. A so-called two-electron model was derived for low resistivity metals [38]. This model yielded

$$\alpha_\lambda(\theta=0,\lambda,T_s) = Af(\lambda/\lambda_{12}) + \frac{B}{C+\lambda^2} \tag{2-85}$$

where

$$f(\lambda/\lambda_{12}) = \left\{\frac{(1+\lambda^2/\lambda_{12}{}^2)^{1/2}-1}{\lambda^2/2\lambda_{12}{}^2}\right\}^{1/2} \tag{2-86}$$

and

$$\lambda_{12}(T_s) = \lambda_o r_e(T_o)/r_e(T_s) \tag{2-87}$$

$$A(T_s) = A_o r_e(T_s)/r_e(T_o) \tag{2-88}$$

$$B(T_s) = B_o-(C+\lambda_x{}^2)A(T_s) \tag{2-89}$$

λ_o, λ_x, C constants

Table 2-2 shows parameters for nickel and platinum [2, 37 - 38], and Table 2-3 shows values for a number of other metals when $T_s \doteq 300$ K. In view of Eqs. (2-77 and 82), the total absorptivity and emissivity values can be obtained approximately from the spectral formula Eq. (2-85) merely by using a mean wavelength λ_e for absorptivity with surrounds at temperature T_e and λ_s for emissivity at temperature T_s.

$$\alpha_T(\theta=0,T_s,T_e) \doteq \alpha_\lambda(\theta=0,\lambda_e,T_s) \tag{2-90}$$

$$\varepsilon_T(\theta=0,T_s) \doteq \alpha_\lambda(\theta=0,\lambda_s,T_s) \quad (2\text{-}91)$$

$$\varepsilon_{T_i}(\theta=0,T_s) \doteq \alpha_\lambda(\theta=0,\lambda_i,T_s) \quad (2\text{-}92)$$

where

$$\lambda_e T_e = \lambda_s T_s \doteq 4050 \ \mu\text{m K}$$

$$\lambda_i T_s \doteq 3180 \ \mu\text{m K}$$

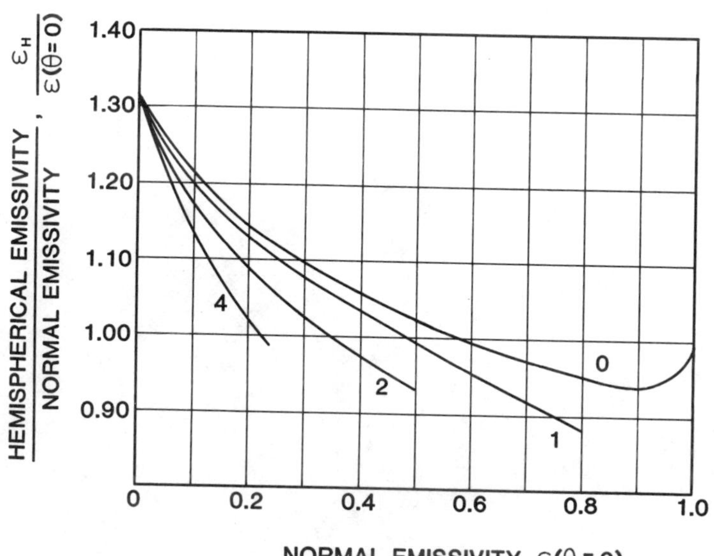

Figure 2-4 Ratio of Hemispherical to Normal Emissivity as a Function of Normal Emissivity and k/n (from R. V. Dunkle [36])

TABLE 2-2

TEMPERATURE DEPENDENCE OF SPECTRAL ABSORPTIVITY AND EMISSIVITY FOR NICKEL AND PLATINUM

(300 K < T < 1400 K)	(1.5 μm < λ < 25 μm)	
PARAMETER	NICKEL	PLATINUM
A_o	0.029	0.034
B_o	1.158 μm^2	0.577 μm^2
C	3.3 μm^2	0.0
T_o	306 K	306 K
n	1.7	0.94
λ_o	5.0 μm	7.0 μm
λ_x	1.5 μm	1.7 μm

TABLE 2-3

SPECTRAL ABSORPTIVITY PARAMETERS FOR METALS AT ROOM TEMPERATURE, FROM REF. [38]

Metal	A	B	C	λ_{12}
Aluminum foil	0.0165	0.23	8.9	14
Cadmium, 99.99% rolled plate	0.054	2.15	3.2	9
Chromium, polished electroplate	0.076	1.58	3.9	3
Columbium, 99.99% rolled plate	0.15	0.29	∿ 0	1
Copper, 99.99%, polished	0.018	0.077	3.2	45
Gold, 99.99%, polished	0.020	0.056	1.4	45
Indium, 99.99%, scraped	0.060	0.24	1.3	6
Inconel X, rolled plate	0.44	0.036	∿ 0	1
Lead, 99.99%, scraped	0.16	0.39	1.1	4
Manganese, 99.99%, polished	0.19	4.8	11	8
Molybdenum, 99.99%	0.033	0.36	∿ 0	7
Nickel, 99.99%, polished	0.029	0.83	2.4	5
Platinum, 99.99%, cold rolled	0.038	0.42	0	4
Rhodium, polished electroplate	0.06	1.27	10	6
Silver, polished electroplate	0.011	0.16	11	70
Stainless steel, 303, lapped	∿0.71	∿0	∿ 0	∿ 0.125
Tin, 99.99%, rolled plate	0.052	0.56	0.8	7
Titanium, polished electroplate	0.13	2.9	8.3	12
Titanium, 99%, lapped	0.09	6.5	15	12
Tungsten, 99.99%, lapped	0.05	0.49	0.3	3
Vanadium, 99.99%, rolled plate	0.17	0.66	0.93	1
Zinc, 99.9%	0.036	0.26	∿ 0	8
Zirconium, 99.99%, rolled plate	0.64	4	35	1

2.H <u>Polarization</u>. The radiant heating of a wall as determined by its absorptivity is seen to be not merely a property of the wall but also the state of polarization of the incident radiation. The engineer can often ignore polarization and still achieve acceptable accuracy in practical situations where the directions of polarization are scrambled in the interreflection process. For example, it was found [41-42] that the transmission of square and circular specular-walled passages could be reasonably well calculated ignoring polarization, but not for the parallel slot. In instruments, such as the integrating sphere and heated cavity reflectometers described earlier, polarization in the instrument optics may be a significant source of error, for angles of incidence considerably off normal.

In Section 4.C a Monte Carlo ray tracing technique is described. If such a computer program is used for radiant heat transfer calculations, one can almost as easily account for polarization as ignore it, thus avoiding having to assess the error in ignoring it. To carry out the ray-tracing calculation, it is necessary for the computer to calculate the fraction of radiation absorbed at a surface from a beam of radiation already polarized by previous reflections or transmissions. The properties of the surface will be calculated from the Fresnel coefficients $\tilde{r}_{p_2}$ and $\tilde{r}_{s_2}$.

The subscript 2 denotes the set of p and s directions defined by the ray (unit vector $\underline{r}$ directed toward the surface) and the surface normal (vector $\underline{n}$).

$$\underline{s}_2 = \frac{\underline{n}x(-\underline{r})}{|\underline{n}x\underline{r}|} \quad , \quad \underline{p}_2 = \underline{s}_2 \times \underline{r} \qquad (2\text{-}93)$$

However, the polarization of the beam is known in terms of the p_1, s_1 directions defined by the plane of incidence of the previous surface. To find I_{p_2} and I_{s_2} it is necessary to resolve the electric vector expressed in the p_1 and s_1 coordinates into the p_2 and s_2 coordinates. With the magnetic vector then fixed from Eq. (2-34), Eq. (2-46) gives I_{p_2} and I_{s_2}. It is found that, in general, insufficient information about the electric field vector is carried by I_{p_1} and I_{s_1}. Additional information is needed regarding the temporally and spectrally smoothed averages of the sine and cosine of the phase relation between the p_1 and s_1 components of the electric vector. Hence two more quantities, e.g. the third and fourth Stokes coefficients, must be introduced.

Consider an approaching wave as described by Eq. (2-24) where coordinate x is in the p_1 direction, y in the s_1 direction, and z along the direction of propagation. Equations (2-33) and (2-46) show that the total intensity I is given by

$$Id\Omega = \frac{1}{4}\,[\underline{k}\cdot\underline{\tilde{E}}x\underline{\tilde{H}}^* + \underline{k}\cdot\underline{\tilde{E}}^*x\underline{\tilde{H}}] = \frac{n}{2\mu_m c_o}\,(\tilde{A}_p\tilde{A}_p^* + \tilde{A}_s\tilde{A}_s^*) \qquad (2\text{-}94)$$

Hence, let

$$\tilde{Z}_p = \left(\frac{n}{2\mu_m c_o d\Omega}\right)^{1/2} \tilde{A}_p \ , \quad \tilde{Z}_s = \left(\frac{n}{2\mu_m c_o d}\right)^{1/2} \tilde{A}_s \qquad (2\text{-}95)$$

In general, there can be an instantaneous phase difference between $\tilde{E}_p$ and $\tilde{E}_s$ or $\tilde{Z}_p$ and $\tilde{Z}_s$, and they can have different instantaneous magnitudes

$$\tilde{Z}_{p_1} = \tilde{M}_{p_1} e^{i\delta_{p_1}} \ , \quad \tilde{Z}_{s_1} = \tilde{M}_{s_1} e^{i\delta_{s_1}} \qquad (2\text{-}95)$$

We thus identify I_{p_1} and I_{s_1} as follows

$$I = I_{p_1} + I_{s_1} \ , \quad I_{p_1} = \langle \tilde{Z}_{p_1} \tilde{Z}^*_{p_1} \rangle \ , \ I_{s_1} = \langle \tilde{Z}_{s_1} \tilde{Z}^*_{s_1} \rangle \qquad (2\text{-}96)$$

where the carets < > denote temporal and spectral smoothing.

Let p_2 and s_2 be directions twisted from the p_1, s_1 directions by angle β.

$$\cos\beta = \underline{p}_2 \cdot \underline{p}_1 \qquad \sin\beta = \underline{p}_2 \cdot \underline{s}_1 \qquad (2\text{-}97)$$

Then

$$\tilde{Z}_{p_2} = \underline{p}_2 \cdot \underline{p}_1 \tilde{Z}_{p_1} + \underline{p}_2 \cdot \underline{s}_1 \tilde{Z}_{s_1} = \cos\beta\, \tilde{Z}_{p_1} + \sin\beta\, \tilde{Z}_{s_1}$$

$$\tilde{Z}_{s_2} = \underline{s}_2 \cdot \underline{p}_1 \tilde{Z}_{p_1} + \underline{s}_2 \cdot \underline{s}_1 \tilde{Z}_{s_1} = -\sin\beta\, \tilde{Z}_{p_1} + \cos\beta\, \tilde{Z}_{s_1}$$

Thus

$$I_{p_2} = \langle \tilde{Z}_{p_2} \tilde{Z}^*_{p_2} \rangle = \langle (\cos\beta\, Z_{p_1} + \sin\beta\, Z_{s_1})(\cos\beta\, Z^*_{p_1} + \sin\beta\, Z^*_{s_1}) \rangle$$

$$I_{p_2} = \cos^2\beta\, I_{p_1} + \sin^2\beta I_{s_1} + \sin\beta\cos\beta \langle \tilde{Z}_{p_1} \tilde{Z}^*_{s_1} + \tilde{Z}^*_{p_1} \tilde{Z}_{s_1} \rangle$$

The quantity needed to find I_{p_2} in addition to I_{p_1} and I_{s_1} is the third Stokes coefficient

$$I_{u_1} = \langle Z_{p_1} Z^*_{s_1} + Z^*_{p_1} Z_{s_1} \rangle = 2R\langle \tilde{Z}_{p_1} \tilde{Z}^*_{s_1} \rangle \tag{2-98}$$

The fourth Stokes coefficient is

$$I_{n_1} = \langle i(\tilde{Z}^*_{p_1} \tilde{Z}_{s_1} - \tilde{Z}_{p_1} \tilde{Z}^*_{s_1} \rangle = 2I\langle \tilde{Z}_{p_1} \tilde{Z}^*_{s_1} \rangle \tag{2-99}$$

The twist transformation of the set of I_{p_1}, I_{s_1}, I_{u_1}, I_{v_1} into I_{p_2}, I_{s_2}, I_{u_2}, I_{v_2} is compactly represented as matrix multiplication

$$\begin{vmatrix} I_{p_2} \\ I_{s_2} \\ I_{u_2} \\ I_{v_2} \end{vmatrix} = \begin{vmatrix} \cos^2\beta & \sin^2\beta & \sin\beta\cos\beta & 0 \\ \sin^2\beta & \cos^2\beta & -\sin\beta\cos\beta & 0 \\ -2\sin\beta\cos\beta & 2\sin\beta\cos\beta & \cos^2\beta-\sin^2\beta & 0 \\ 0 & 0 & 0 & 1 \end{vmatrix} \cdot \begin{vmatrix} I_{p_1} \\ I_{s_1} \\ I_{u_1} \\ I_{v_1} \end{vmatrix} \tag{2-100}$$

Due to the random nature of black body radiation one has $\langle M_p \rangle = \langle M_s \rangle$ and $\langle \cos(\delta_p - \delta_s) \rangle = \langle \sin(\delta_p - \delta_s) \rangle = 0$. Thus for black body radiation $\langle \tilde{Z}_p \tilde{Z}^*_p \rangle = \langle \tilde{Z}_s \tilde{Z}^*_s \rangle = 0$ and $\langle z_p z^*_s \rangle = \langle z_s z^*_p \rangle = 0$, i.e.,

$$I_{b,p} = I_{b,s} = I_b/2 \;, \quad I_{b,u} = I_{b,v} = 0 \tag{2-101}$$

A column vector carries the requisite information

$$|I_b| = |X_b| I_b \tag{2-102}$$

$$|X_b| = \begin{vmatrix} 1/2 \\ 1/2 \\ 0 \\ 0 \end{vmatrix} \tag{2-103}$$

Finding the intensity I can be represented by multiplying

by a row matrix, designated with a super t (for "transpose"):

$$|E|^t = |1\ ,\ 1\ ,\ 0\ ,\ 0| \tag{2-104}$$

$$I = |E|^t \cdot |I| \tag{2-105}$$

With the intensity matrix components known in terms of the principal p_2, s_2 coordinates, the Fresnel coefficients can be used to find the intensity matrix of the reflected ray. Let a prime on the subscript denote the reflected ray and drop the double subscript 2 to simplify the notation. Then

$$\tilde{Z}_{p'} = \tilde{r}_p \tilde{Z}_p \qquad \tilde{Z}_s = \tilde{r}_s \tilde{Z}_s \tag{2-106}$$

Hence

$$I_{p'} = \langle \tilde{r}_p \tilde{Z}_p \tilde{r}_p^* \tilde{Z}_p^* \rangle = \tilde{r}_p \tilde{r}_p^* \langle \tilde{Z}_p \tilde{Z}_p^* \rangle = \tilde{r}_p \tilde{r}_p^* I_p$$

$$I_{u'} = 2\mathcal{R} \langle \tilde{r}_p \tilde{Z}_p \tilde{r}_s^* \tilde{Z}_s^* \rangle$$

$$I_{u'} = \mathcal{R}\{\tilde{r}_p \tilde{r}_s^*\} \mathcal{R} \langle 2\tilde{Z}_p \tilde{Z}_s^* \rangle - \mathcal{I}\{\tilde{r}_p \tilde{r}_s^*\} \mathcal{I} \langle 2\tilde{Z}_p \tilde{Z}_s^* \rangle$$

$$I_{u'} = \mathcal{R}\{\tilde{r}_p \tilde{r}_s^*\} I_u - \mathcal{I}\{\tilde{r}_p \tilde{r}_s^*\} I_v$$

Similarly the $I_{s'}$ and $I_{v'}$ quantities can be found. The column matrix for the reflected intensity is thus

$$\begin{vmatrix} I_{p'} \\ I_{s'} \\ I_{u'} \\ I_{v'} \end{vmatrix} = \begin{vmatrix} \rho_{pp} & 0 & 0 & 0 \\ 0 & \rho_{ss} & 0 & 0 \\ 0 & 0 & \rho_{uu} & \rho_{uv} \\ 0 & 0 & \rho_{vu} & \rho_{vv} \end{vmatrix} \cdot \begin{vmatrix} I_p \\ I_s \\ I_u \\ I_v \end{vmatrix} \tag{2-107}$$

where

$$\rho_{pp} = \tilde{r}_p\tilde{r}_p^* , \quad \rho_{ss} = \tilde{r}_s\tilde{r}_s^*$$

$$\rho_{uu} = \rho_{vv} = \mathcal{R}\{\tilde{r}_p\tilde{r}_s^*\} , \quad \rho_{uv} = -\rho_{vu} = -\mathcal{I}\{\tilde{r}_p\tilde{r}_s^*\} \qquad (2\text{-}108)$$

The fraction of the incident energy reflected is then

$$\rho = \frac{|E|^t \cdot |\rho| \cdot |I|}{|E|^t \cdot |I|} = \frac{|E|^t \cdot |\rho| \cdot |X|}{|E|^t \cdot |X|} \qquad (2\text{-}109)$$

and the fraction absorbed is $1-\rho$. It is these fractions that are needed for a ray tracing calculation.

The state of polarization of the leaving ray is contained in the column matrix $|X'|$

$$|X'| = \frac{|I'|}{|E^t| \cdot |I'|} = \frac{|\rho| \cdot |X|}{|E^t| \cdot |\rho| \cdot |X|} \qquad (2\text{-}110)$$

The p'-s' coordinates and the direction of the reflected ray are also needed. The emergent ray r' maintains its component tangent to the surface in the plane of incidence and reverses its component normal to the surface

$$\underline{r}' = -[\underline{r}\cdot\underline{n}]\underline{n} + [\underline{r}\cdot(\underline{n}x\underline{s})](\underline{n}x\underline{s}) \qquad (2\text{-}111)$$

$$\underline{s}' = \underline{s} \qquad (2\text{-}112)$$

$$\underline{p}' = \underline{s}'x\underline{r}' \qquad (2\text{-}113)$$

With this information the ray tracing procedure of Section 4.C can continue to the next wall encounter. Drop the prime on $\underline{r}'$, take the normal to the next wall $\underline{n}$ and use Eq. (2-93) to find the new p_2-s_2 directions. Taking p'-s' to be p_1-s_1, one finds the twist angle β from Eq. (2-97) and transforms $|X'|$ to $|X|$ using Eq. (2-100). Then the reflectivity can be found from Eq. (2-109), and the ray tracing can be continued by using Eqs. (2-110) to (2-113) and repeating the entire process.

As stated at the outset of this section, the designer can often ignore polarization except for special circumstances where several interreflections occur with little or no β-twist or where instruments have polarizing optics. In Section 2.I material selection where polarization is not a significant factor is briefly discussed. In the whole of Section 3, where the surfaces are assumed to be perfectly diffuse, and are thus depolarizing by definition, polarization is not a concern. The phenomenon impacts Section 4, where radiant interchange in the presence of specular surfaces is treated. Polarization is also an important feature in scattering from clouds of particles, one which aids inferring particle sizes.

2.I Thermal Radiation Characteristics in Thermal Design. The designer generally wants either to promote or suppress thermal radiation in a given spectral region. The spectral requirements are sometimes different in different spectral regions. If the spectral regions are classified crudely as long wavelengths and short wavelengths, there are four combinations of surface properties, one of which will be often chosen as the main surface finish. Occasionally one of the others will be used in a striped or chessboard pattern to balance the design requirements. The four basic finishes are illustrated in Fig. 2.5. The flat black absorber is achieved with black paint, black glass enamel, and heavy black oxide or dyed anodized coatings, as shown in Figs. 2-6a, 2-7, and 2-8. It is used where high emissivity is desired in the long wavelengths and high solar absorptivity in the short wavelengths is no handicap. The flat reflector is achieved with bare metals and leafing aluminum paint, as may be seen from inspection of Table 2-4. The metal is at times applied by electroplating or vacuum evaporation. If high specularity is required, a glossy enamel or plastic subcoating may be applied first. For high temperature applications or corrosive environments, layers to act as barriers to foreign metal and oxygen diffusion may be employed as under and/or over coats. Aluminum foil and aluminized plastic film are used for low temperatures. The flat reflector provides a low emissivity and high reflectivity for both long and short wavelengths. Because the characteristics are the same in the long and short wavelengths, both the flat absorber and flat reflector are said to be nonselective.

The selective black solar absorber has a high absorptivity in the short wavelengths and simultaneously a low emissivity in the long wavelengths. A metal surface with semiconductor coating, particularly a thin one with a gradient in the index of refraction, serves. Thin copper or nickel or chrome oxide layers formed by chemical dip or electrochemical finishing are used on copper plated steel, as shown in Fig. 2-9. Spray-on and baked finishes can be applied to aluminum, and vacuum deposited films are used. Such surfaces are used for solar heat collectors or to maintain warm temperatures for space vehicles exploring the outer reaches of the solar system.

The solar reflector has a high reflectivity in the short wavelengths and a high emissivity in the long wavelengths. White paint, clear lacquered aluminum, and soft-anodized aluminum, second-surface mirrors flame or plasma sprayed ceramic coatings, ceramics, and fired enamels are examples in Figs. 2-6b, 2-6c, 2-8b, 2-10, and 2.11. The finish is used to emit unwanted heat and simultaneously reflect unwanted solar energy. The tops of vehicles and structures and instrument cases exposed to solar heating are often painted white.

FIGURE 2-5

Surface Finishes Often Used in Thermal Design
Ref. [44]

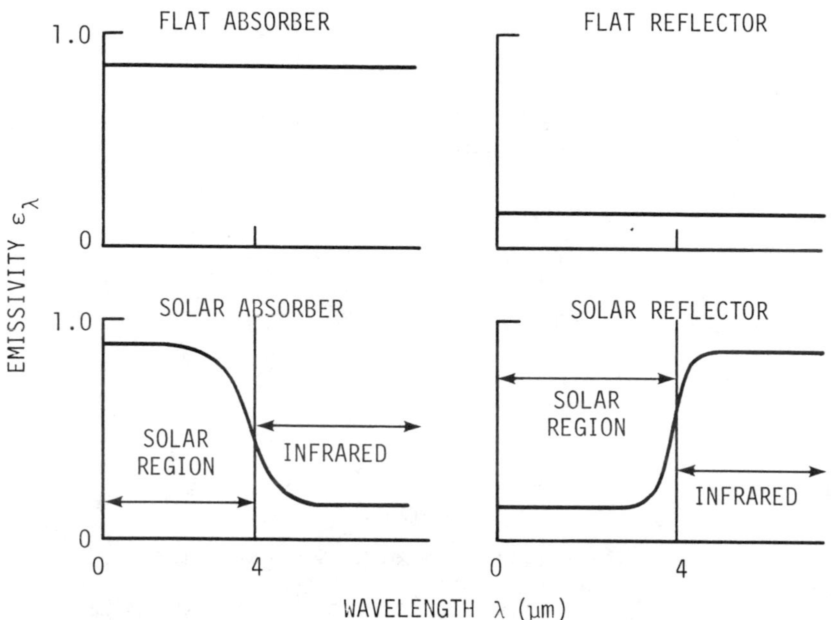

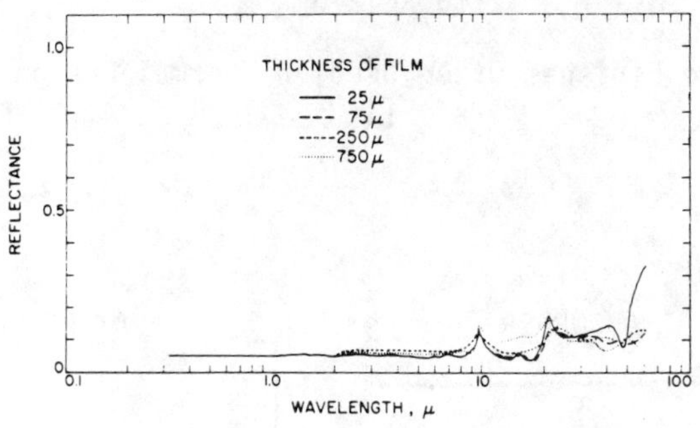

Reflectance of Cat-a-lac epoxy black paint on aluminum substrate.

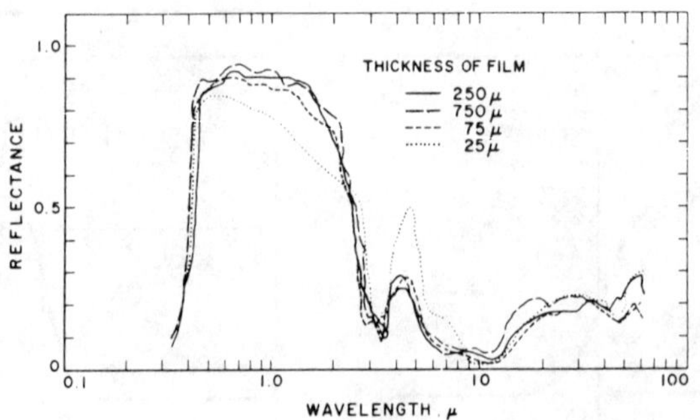

Reflectance of PV-100 silicone-alkyd white paint on aluminum substrate.

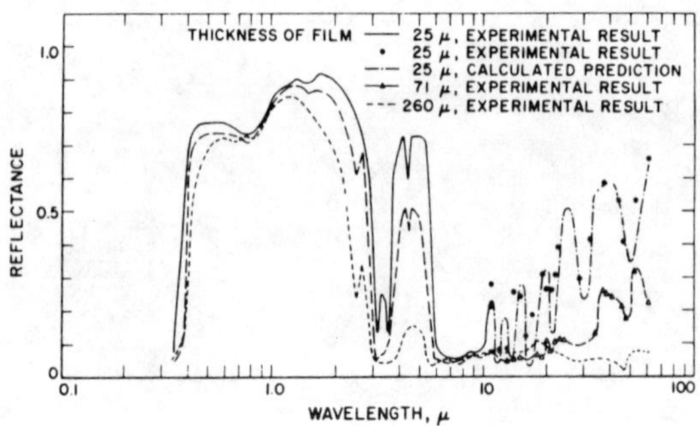

Reflectance of Laminar X-500 polyurethane clear film on aluminum substrate.

Figure 2-6a,b,c Spectral Directional Reflectance ($\theta \doteq 0$) for a Black, White, and Clear Paint (From Ref. [47])

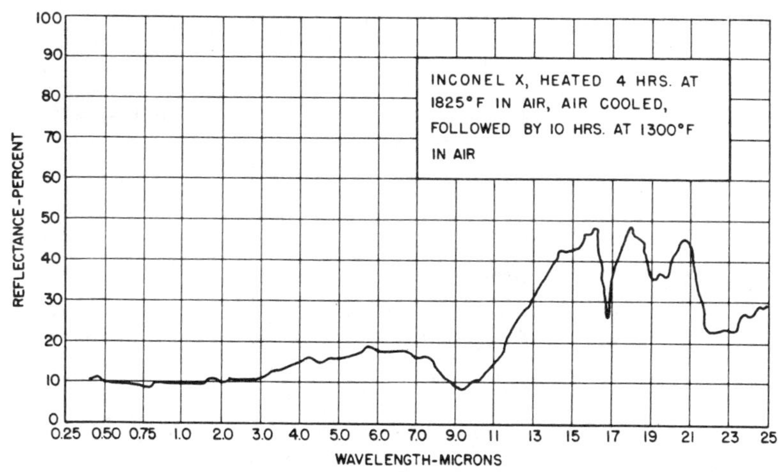

Near-normal spectral reflectance of oxidized Inconel X

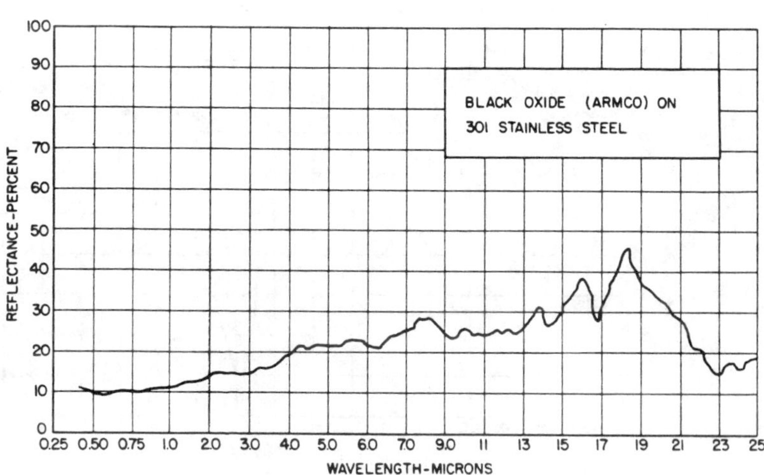

Near-normal spectral reflectance of oxidized stainless steel

Figure 2-7a,b Spectral Directional Reflectance ($\theta \doteq 0$) of Two Black Oxidized Metals (From Ref. [46]. Note change of Wavelength Scale at 1 and at 7 μm)

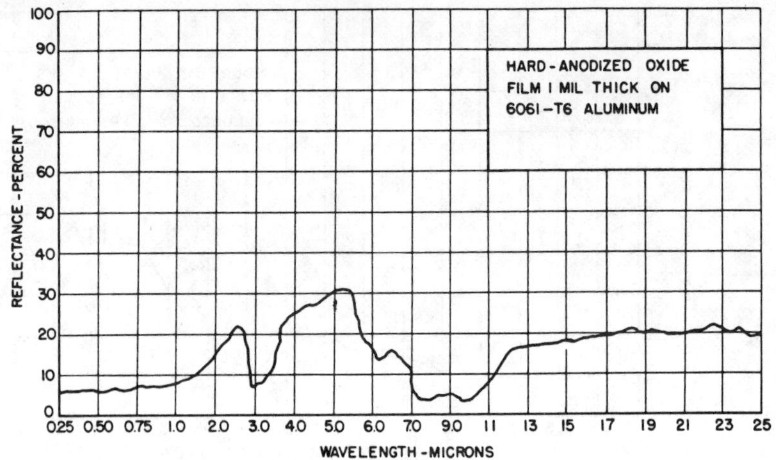

Near-normal spectral reflectance of hard-anodized aluminum

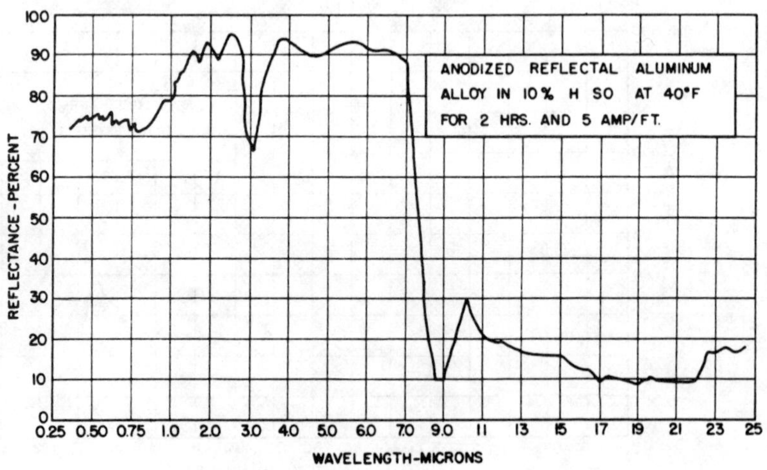

Near-normal spectral reflectance of soft-anodized aluminum

Figure 2-8a,b Spectral Directional Reflectance ($\theta \doteq 0$) of Hard and Soft Anodized Aluminum (From Ref. [46]. Note change of wavelength scale at 1 and at 7 μm)

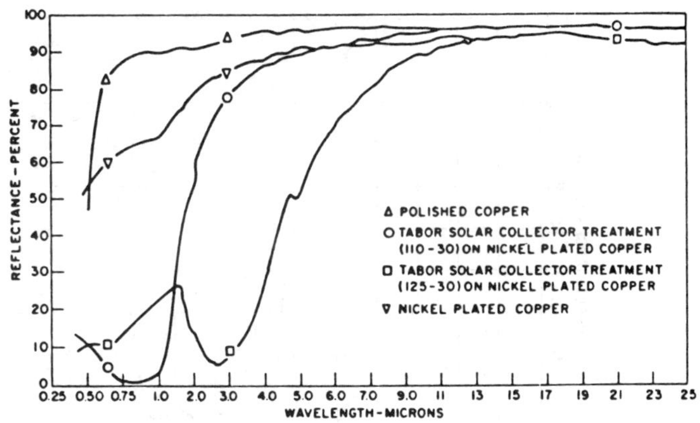

Spectral Reflectance of Tabor Selective Blacks on Copper

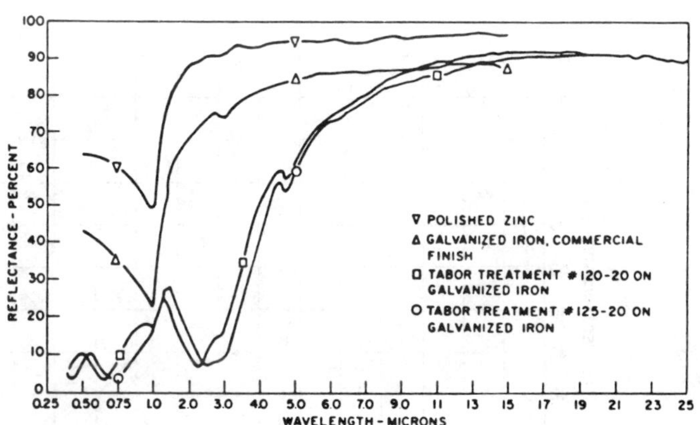

Spectral Reflectance of Tabor Selective Blacks on Zinc

Figure 2-9a,b Spectral Directional Reflectance ($\theta \doteq 0$) of Some Selective Black Surfaces (From Ref. [51]. Note change of wavelength scale at 1 and at 7 μm)

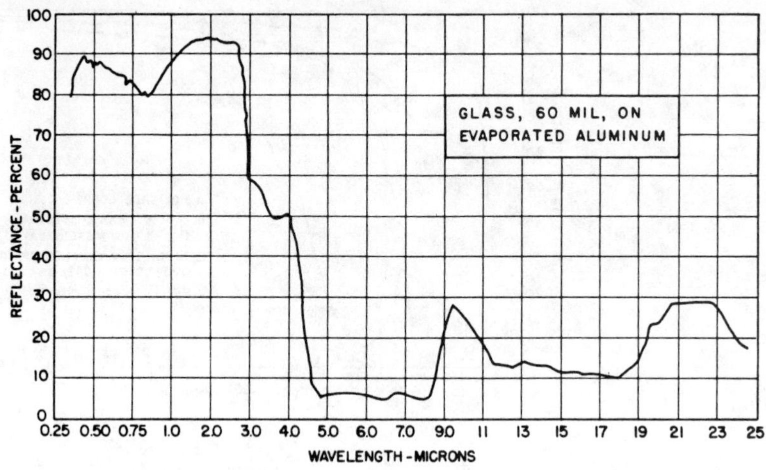

Near-normal spectral reflectance of glass on evaporated aluminum

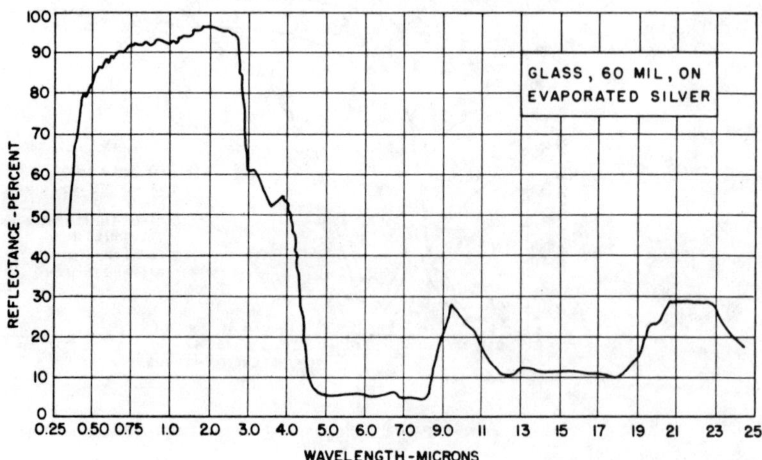

Near-normal spectral reflectance of glass on evaporated silver

Figure 2-10a,b Spectral Directional Reflectance ($\theta \doteq 0$) of Two Second-Surface Mirrors (From Ref. [46]. Note change of wavelength scale at 1 and at 7 μm)

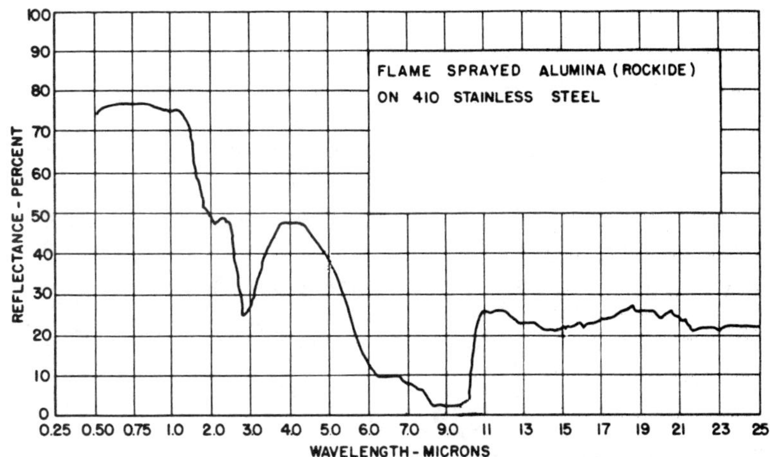

Near-normal spectral reflectance of flame-sprayed alumina

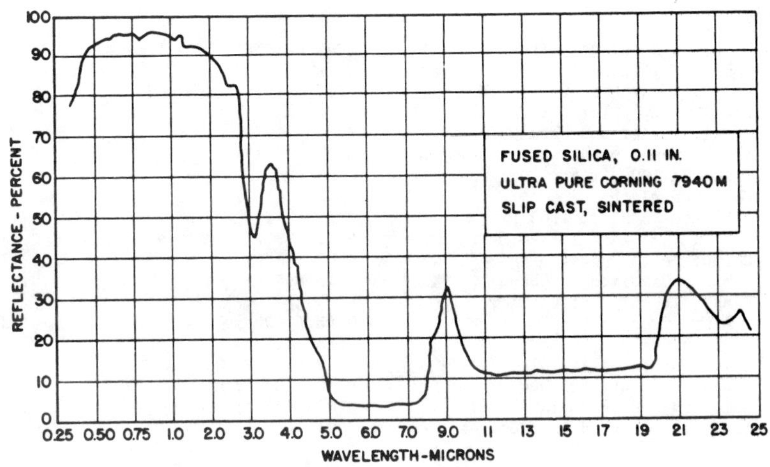

Near-normal spectral reflectance of fused silica

Figure 2-11a,b Spectral Directional Reflectance ($\theta \doteq 0$) of Alumina and Silica Ceramics (From Ref. [46]. Note change of wavelength scale at 1 and at 7 μm)

TABLE 2-4

TOTAL EMISSIVITY AND SOLAR ABSORPTIVITY OF SELECTED SURFACES, A COMPILATION FROM REF. [52]

	Total Normal Emittance[a]	Extraterrestrial Solar Absorptance
Alumina, flame sprayed	0.80 (-10°F)	0.28
Aluminum foil, as received	0.04 (70°F)	
Aluminum foil, bright dipped	0.025 (70°F)	0.10
Aluminum, vacuum deposited on duPont mylar	0.025 (70°F)	0.10
Aluminum alloy, 6061, as received	0.03 (70°F)	0.37
Aluminum alloy, 75S-T6 weathered 20,000 hr on a DC6 aircraft	0.16 (150°F)	0.54
Aluminum, hard anodized, 6061-T6, 35 amp/ft^2 at 45 volts in 20°F sulfuric acid solution, 1 mil thick	0.84 (-10°F)	0.92
Aluminum, soft anodized *Reflectal* aluminum alloy, 5 amp/ft^2 for 2 hr in 40°F, 10% H_2SO_4 solution	0.79 (-10°F)	0.23
Aluminum, 7075-T6, sandblasted with 60 mesh silicon carbide grit	0.30 (70°F)	0.55
Aluminized silicone resin paint	0.20 (200°F)	0.27
Dow Corning XP-310	0.22 (800°F)	
Beryllium	0.18 (300°F)	0.77
	0.21 (700°F)	
	0.30 (1100°F)	
Beryllium, anodized	0.90 (300°F)	
	0.88 (700°F)	
	0.82 (1100°F)	
Black paint		
Parson's optical black	0.95 (-10°F)	0.975
Black silicone high heat National Lead Co. 46H47	0.93 (-10°F to 1400°F)	0.94
Black epoxy paint, cat-a-lac Finch Paint and Chemical Co. 463-1-8	0.89 (-10°F)	0.95
Black enamel paint, Rinshed-Mason	0.81 (200°F)	
heated 1000 hr at 700°F in air	0.80 (800°F)	
Chromium plate	0.12 (200°F)	
heated 50 hr at 1100°F	0.15 (750°F)	
	0.15 (95°F)	0.78
Copper, electroplated and	0.03 (70°F)	0.47
black oxidized in Ebonol C	0.16 (95°F)	0.91
Glass, second surface mirror		
Aluminized	0.83 (-10°F)	0.13
Silvered	0.83 (-10°F)	0.13
Gold, coated on stainless steel,	0.09 (200°F)	
heated in air at 1000°F	0.14 (750°F)	
Gold, coated on 3M Tape Y9814	0.025 (70°F)	0.21
Graphite, crushed on sodium silicate	0.91 (-10°F)	0.96

Table 2-4 (continued)

	Total Normal Emittance	Extraterrestrial Solar Absorptance
Inconel X, oxidized 4 hr at 1825°F	0.71 (-10°F)	0.90
10 hr at 1300°F in air	0.81 (200°F)	
	0.79 (800°F)	
Magnesium-thorium alloy	0.07 (200°F)	
	0.06 (500°F)	
Magnesium, Dow 7 coating	0.36 (700°F)	
Mylar, duPont film aluminized on second surface		
0.00025 in. thick	0.37 (70°F)	0.17
0.001 in. thick	0.63 (70°F)	0.17
0.003 in. thick	0.81 (70°F)	0.24
Nickel, electroplated	0.03 (70°F)	0.22
Nickel, Tabor solar absorber electro-oxidized nickel on copper		
110-30	0.05 (95°F)	0.85
125-30	0.11 (95°F)	0.85
Platinum-coated stainless steel	0.13 (200°F)	
	0.15 (750°F)	
Annealed in air 300 hr at 700°F	0.11 (200°F)	
	0.13 (800°F)	
Silica, Corning Glass 7940M sintered powdered fused silica	0.84 (95°F)	0.08
Silica, second surface mirror		
Aluminized	0.83 (70°F)	0.14
Silvered	0.83 (70°F)	0.07
Silicon solar cell		
boron doped, no coverglass	0.32 (95°F)	0.94
0.006 mil quartz cover, blue filter	0.81 (70°F)	0.81
Silver		
Plated on nickel on stainless steel	0.06 (200°F)	
	0.08 (750°F)	
Heated 300 hr at 700°F	0.11 (200°F)	
	0.13 (800°F)	
Silver Chromatone paint	0.24 (70°F)	0.20
Stainless steel		
Type 312 heated 300 hr at 500°F	0.27 (200°F)	
	0.32 (800°F)	
Type 301 with Armco black oxide	0.75 (-10°F)	0.89
Type 410 heated to 1300°F in air	0.13 (95°F)	0.76
Type 303 sandblasted heavily with 80 mesh aluminum oxide grit	0.42 (200°F)	0.68
Steel blackened by Ebonol S treatment dipped 15 min in 286°F boiling solution	0.10 (95°F)	0.85
Titanium		
75A	0.10 (200°F)	
	0.19 (800°F)	
75A oxidized 300 hr at 850°F in air	0.21 (95°F)	0.80
	0.25 (800°F)	

Table 2-4 (continued)

	Total Normal Emittance	Extraterrestrial Solar Absorptance
C-110M oxidized 100 hr at 800°F in air	0.16 (95°F)	0.52
C-110M oxidized 300 hr at 850°F in air	0.20 (95°F)	0.77
Evaporated 80 to 100 μm thick and oxidized 3 hr at 750°F in air	0.14 (95°F)	0.75
Anodized	0.73 (-10°F)	0.51
White acrylic resin paint	0.92 (200°F)	
Sherwin-Williams M49WC8-CA-10144	0.87 (400°F)	
White epoxy paint, Cat-a-lac Finch Paint and Chemical Co. 483-1-8	0.88 (-10°F)	0.25
White potassium zirconium silicate inorganic spacecraft coating	0.89 (70°F)	0.13
Zinc blackened by Tabor solar collector electrochemical treatment 120-20	0.12 (95°F)	0.89

[a] For high emittance materials these values are about 0.03 higher than hemispherical values, while for low emittance metals the values are about 25% lower than hemispherical emittance. See, for example, *Progress in Astronautics and Aeronautics*, Vol. 11, pp. 427-446, 1963; *Solar Energy*, Vol. VI, No. 1, pp. 1-8, 1962; and Report 60-93 of the Department of Engineering, UCLA, 1960.

TABLE 2-5

TOTAL NORMAL EMISSIVITY OF VARIOUS SURFACES,
A COMPILATION FROM REF. [53]

	T, °F[a]	Emissivity
A. Metals and Their Oxides		
Aluminum:		
Highly polished plate, 98.3%, pure	440-1070	0.039-0.057
Commercial sheet	212	0.09
Rough polish	212	0.18
Rough plate	100	0.055-0.07
Oxidized at 1110°F	390-1110	0.11-0.19
Heavily oxidized	200-940	0.20-0.31
Aluminum oxide	530-930	0.63-0.42
Aluminum oxide	930-1520	0.42-0.26
Al-surfaced roofing	100	0.216
Aluminum alloys[b]		
Alloy 75 ST; A, B_1, C	75	0.11,0.10,0.08
Alloy 75 ST; A[c]	450-900	0.22-0.16
Alloy 75 ST; B_1[c]	450-800	0.20-0.18
Alloy 75 ST; C[c]	450-930	0.22-0.15
Alloy 24 ST; A, B_1, C	75	0.09
Alloy 24 ST; A[c]	450-910	0.17-0.15
Alloy 24 ST; B_1[c]	450-940	0.20-0.16
Alloy 24 ST; C[c]	450-860	0.16-0.13
Calorized surfaces, heated at 1110°F		
Copper	390-1110	0.18-0.19
Steel	390-1110	0.52-0.57
Brass:		
Highly polished		
73.2 Cu, 26.7 Zn	476-674	0.028-0.031
62.4 Cu, 36.8 Zn, 0.4 Pb, 0.3 Al	494-710	0.033-0.037
82.9 Cu, 17.0 Zn	530	0.030
Hard-rolled, polished, but direction of polishing visible	70	0.038
Hard-rolled, polished, but somewhat attacked	73	0.043
Hard-rolled, polished, but traces of stearin from polish left on	75	0.053
Polished	212	0.06
Polished	100-600	0.10
Rolled plate, natural surface	72	0.06
Rolled plate, rubbed with coarse emery	72	0.20
Dull plate	120-660	0.22
Oxidized by heating at 1110°F	390-1110	0.61-0.59
Chromium (see nickel alloys for Ni-Cr steels):		
Polished	100-2000	0.08-0.36
Polished	212	0.075

Table 2-5 (continued)

	T, °F[a]	Emissivity
Copper:		
Carefully polished electrolytic copper	176	0.018
Polished	242	0.023
Polished	212	0.052
Commercial emeried, polished, but pits remaining	66	0.030
Commercial, scraped shiny, but not mirror-like	72	0.072
Plate, heated long time, covered with thick oxide layer	77	0.78
Plate heated at 1110°F	390-1110	0.57
Cuprous oxide	1470-2010	0.66-0.54
Molten copper	1970-2330	0.16-0.13
Dow metal:[b]		
A; B_1; C	75	0.15,0.15,0.12
A[c]	450-750	0.24-0.20
B_1[c]	450-800	0.16
C[c]	450-760	0.21-0.18
Gold:		
Pure, highly polished	440-1160	0.018-0.035
Inconel:[b]		
Types X and B; surface A, B_2, C	75	0.19-0.21
Type X; surface A[c]	450-1620	0.55-0.78
Type X; surface B_2[c]	450-1575	0.60-0.75
Type X; surface C[c]	450-1650	0.62-0.73
Type B; surface A[c]	450-1620	0.35-0.55
Type B; surface B_2[c]	450-1740	0.32-0.51
Type B; surface C[c]	450-1830	0.35-0.40
Iron and steel (not including stainless):		
Metallic surfaces (or very thin oxide layer)		
Electrolytic iron, highly polished	350-440	0.052-0.064
Steel, polished	212	0.066
Iron, polished	800-1880	0.14-0.38
Iron, roughly polished	212	0.17
Iron, freshly emeried	68	0.24
Cast iron, polished	392	0.21
Cast iron, newly turned	72	0.44
Cast iron, turned and heated	1620-1810	0.60-0.70
Wrought iron, highly polished	100-480	0.28
Polished steel casting	1420-1900	0.52-0.56
Ground sheet steel	1720-2010	0.55-0.61
Smooth sheet iron	1650-1900	0.55-0.60
Mild steel;[b] A, B_2, C	75	0.12,0.15,0.10
Mild steel;[b] A[c]	450-1950	0.20-0.32
Mild steel;[b] B_2[c]	450-1920	0.34-0.35
Mild steel;[b] C[c]	450-1950	0.27-0.31

Table 2-5 (continued)

	T, °F[a]	Emissivity
Oxidized surfaces		
Iron plate, pickled, then rusted red	68	0.61
Iron plate, completely rusted	67	0.69
Iron, dark gray surface	212	0.31
Rolled sheet steel	70	0.66
Oxidized iron	212	0.74
Cast iron, oxidized at 1110°F	390-1110	0.64-0.78
Steel, oxidized at 1110°F	390-1110	0.79
Smooth oxidized electrolytic iron	260-980	0.78-0.82
Iron oxide	930-2190	0.85-0.89
Rough ingot iron	1700-2040	0.87-0.95
Sheet steel		
Strong, rough oxide layer	75	0.80
Dense, shiny oxide layer	75	0.82
Cast plate, smooth	73	0.80
Cast plate, rough	73	0.82
Cast iron, rough, strongly oxidized	100-480	0.95
Wrought iron, dull oxidized	70-680	0.94
Steel plate, rough	100-700	0.94-0.97
Molten surfaces		
Cast iron	2370-2550	0.29
Mild steel	2910-3270	0.28
Steel, several different kinds with 0.25-1.2% C (slightly oxidized surface)	2840-3110	0.27-0.39
Steel	2730-3000	0.42-0.53
Steel	2770-3000	0.43-0.40
Pure iron	2760-3220	0.42-0.45
Armco iron	2770-3070	0.40-0.41
Lead:		
Pure (99.96%), unoxidized	260-440	0.057-0.075
Gray oxidized	75	0.28
Oxidized at 300°F	390	0.63
Magnesium:		
Magnesium oxide	530-1520	0.55-0.20
Magnesium oxide	1650-3100	0.20
Mercury	32-212	0.09-0.12
Molybdenum:		
Filament	1340-4700	0.096-0.202
Massive, polished	212	0.071
Monel metal:[b]		
Oxidized at 1110°F	390-1110	0.41-0.46
K Monel 5700; A, B_2, C	75	0.23,0.17,0.14
K Monel 5700; A[c]	450-1610	0.46-0.65
K Monel 5700; B_2[c]	450-1750	0.54-0.77
K Monel 5700; C[c]	450-1785	0.35-0.53

Table 2-5 (continued)

	T, °F[a]	Emissivity
Nickel:		
Electroplated, polished	74	0.045
Technically pure (98.9% Ni, + Mn), polished	440-710	0.07-0.087
Polished	212	0.072
Electroplated, not polished	68	0.11
Wire	368-1844	0.096-0.186
Plate, oxidized by heating at 1110°F	390-1110	0.37-0.48
Nickel oxide	1200-2290	0.59-0.86
Nickel alloys:		
Chromnickel	125-1894	0.64-0.76
Copper-nickel, polished	212	0.059
Nichrome wire, bright	120-1830	0.65-0.79
Nichrome wire, oxidized	120-930	0.95-0.98
Nickel-silver, polished	212	0.135
Nickelin (18-32 Ni; 55-68 Cu; 20 Zn), gray oxidized	70	0.262
Type ACI-HW (60 Ni; 12 Cr)		
Smooth, black, firm adhesive oxide coat from service	520-1045	0.89-0.82
Platinum:		
Pure, polished plate	440-1160	0.054-0.104
Strip	1700-2960	0.12-0.17
Filament	80-2240	0.036-0.192
Wire	440-2510	0.073-0.182
Silver:		
Polished, pure	440-1160	0.020-0.032
Polished	100-700	0.022-0.031
Polished	212	0.052
Stainless steel:[b]		
Polished	212	0.074
Type 301; A, B_2, C	75	0.21,0.27,0.16
Type 301; A[c]	450-1740	0.57-0.55
Type 301; B_2[c]	450-1725	0.54-0.63
Type 301; C[c]	450-1650	0.51-0.70
Type 316; A, B_2, C	75	0.28,0.28,0.17
Type 316; A[c]	450-1600	0.57-0.66
Type 316; B_2[c]	450-1920	0.52-0.50
Type 316; C[c]	450-1920	0.26-0.31
Type 347; A, B_2, C	75	0.39,0.35,0.17
Type 347; A[c]	450-1650	0.52-0.65
Type 347; B_2[c]	450-1610	0.51-0.65
Type 347; C[c]	450-1650	0.49-0.64
Type 304; (8 Cr; 18 Ni)		
Light silvery, rough, brown, after heating	420-914	0.44-0.36
After 42 hr heating at 980°F	420-980	0.62-0.73

Table 2-5 (continued)

	T, °F[a]	Emissivity
Type 310 (25 Cr; 20 Ni)		
Brown, splotched, oxidized from furnace service	420-980	0.90-0.97
Allegheny metal No. 4, polished	212	0.13
Allegheny alloy No. 66, polished	212	0.11
Tantalum filament	2420-5430	0.19-0.31
Thorium oxide	530-930	0.58-0.36
Thorium oxide	930-1520	0.36-0.21
Tin:		
Bright tinned iron	76	0.043 and 0.064
Bright	122	0.06
Commercial tin-plated sheet iron	212	0.07,0.08
Tungsten:		
Filament, aged	80-6000	0.032-0.35
Filament	6000	0.39
Polished coat	212	0.066
Zinc:		
Commercial 99.1% pure, polished	440-620	0.045-0.053
Oxidized by heating at 750°F	750	0.11
Galvanized sheet iron, fairly bright	82	0.23
Galvanized sheet iron, gray oxidized	75	0.28
Zinc, galvanized sheet	212	0.21
B. Refractories, Building Materials, Paints, and Miscellaneous		
Alumina (99.5-85 Al_2O_3; 0-12 SiO_2; 0-1 Fe_2O_3).		
Effect of mean grain size, microns (μ)	1850-2850	
10 μ		0.30-0.18
50 μ		0.39-0.28
100 μ		0.50-0.40
Alumina-silica (showing effect of Fe)	1850-2850	
80-58 Al_2O_3; 16-38 SiO_2; 0.4 Fe_2O_3		0.61-0.43
36-26 Al_2O_3; 50-60 SiO_2; 1.7 Fe_2O_3		0.73-0.62
61 Al_2O_3; 35 SiO_2; 2.9 Fe_2O_3		0.78-0.68
Asbestos:		
Board	74	0.96
Paper	100-700	0.93-0.94
Brick[d]		
Red, rough, but no gross irregularities	70	0.93
Grog brick, glazed	2012	0.75
Building	1832	0.45
Fireclay	1832	0.75
Carbon:		
T-carbon (Gebrüder Siemens) 0.9% ash. This started with emissivity at 260°F of 0.72, but on heating changed to values given	260-1160	0.81-0.79

Table 2-5 (continued)

	T, °F[a]	Emissivity
Filament	1900-2560	0.526
Rough plate	212-608	0.77
Rough plate	608-932	0.77-0.72
Graphitized	212-608	0.76-0.75
Graphitized	608-932	0.75-0.71
Candle soot	206-520	0.952
Lampblack-waterglass coating	209-440	0.96-0.95
Thin layer of same on iron plate	69	0.927
Thick coat of same	68	0.967
Lampblack, 0.003 in. or thicker	100-700	0.945
Lampblack, rough deposit	212-932	0.84-0.78
Lampblack, other blacks	122-1832	0.96
Graphite, pressed, filed surface	480-950	0.98
Carborundum (87 SiC; density 2.3)	1850-2550	0.92-0.82
Concrete tiles	1832	0.63
Enamel, white fused, on iron	66	0.90
Glass:		
Smooth	72	0.94
Pyrex, lead, and soda	500-1000	Ca 0.95-0.85
Gypsum, 0.02 in. thick on smooth or blackened plate	70	0.903
Magnesite refractory brick	1832	0.38
Marble, light gray, polished	72	0.93
Oak, planed	70	0.90
Oil layers on polished nickel (lubricating oil).	68	
Polished surface, alone		0.045
+0.001, 0.002, 0.005 in oil		0.27,0.46,0.72
Thick oil layer		0.82
Oil layers on aluminum foil (linseed oil)		
Aluminum foil	212	0.087
+1, 2 coats oil	212	0.561,0.574
Paints, lacquers, varnishes:		
Snow-white enamel varnish on rough iron plate	73	0.906
Black shiny lacquer, sprayed on iron	76	0.875
Black shiny shellac on tinned iron sheet	70	0.821
Black matte shellac	170-295	0.91
Black or white lacquer	100-200	0.80-0.95
Flat black lacquer	100-200	0.96-0.98
Oil paints, 16 different, all colors	212	0.92-0.96
Aluminum paints and lacquers:		
10% Al, 22% lacquer body, on rough or smooth surface	212	0.52
Other Al paints, varying age and Al content	212	0.27-0.67
Al lacquer, varnish binder, on rough plate	70	0.39

Table 2-5 (continued)

	T, °F[a]	Emissivity
Al paint, after heating at 620°F	300-600	0.35
Radiator paint; white, cream, bleach	212	0.79,0.77,0.84
Radiator paint; bronze	212	0.51
Lacquer coatings, 0.001-0.015 in. thick on aluminum alloys[e]	100-300	0.87 to 0.97
Clear silicone vehicle coatings, 0.001-0.015 in. thick[e]		
On mild steel	500	0.66
On stainless steels, 316, 301, 347	500	0.68,0.75,0.75
On Dow metal	500	0.74
On Al alloys 24 ST, 75 ST	500	0.77,0.82
Aluminum paint with silicone vehicle, two coats on Inconel[e]	500	0.29
Paper, thin, pasted on tinned or blackened plate	66	0.92,0.94
Plaster, rough lime	50-190	0.91
Porcelain, glazed	72	0.92
Quartz:		
Rough, fused	70	0.93
Glass, 1.98 mm thick	540-1540	0.90-0.41
Glass, 6.88 mm thick	540-1540	0.93-0.47
Opaque	570-1540	0.92-0.68
Roofing paper	69	0.91
Rubber:		
Hard, glossy plate	74	0.94
Soft, gray, rough (reclaimed)	76	0.86
Serpentine, polished	74	0.90
Silica (98 SiO_2; Fe-free), effect of grain size, microns (μ)	1850-2850	
10 μ		0.42-0.33
70-600 μ		0.62-0.46
(See also Alumina-silica and quartz)		
Water	32-212	0.95-0.963
Zirconium silicate	460-930	0.92-0.80
Zirconium silicate	930-1530	0.80-0.52

[a]When temperatures and emissivities appear in pairs separated by dashed, they correspond; and linear interpolation is permissible.

[b]Identification of surface treatment; surface A, cleaned with toluene, then methanol; B_1, cleaned with soap and water, toluene, and methanol in succes sion; B_2, cleaned with abrasive soap and water, toluene, and methanol. C, polished on buffing wheel to mirror surface, cleaned with soap and water

[c]Results after repeated heating and cooling.

[d]See also under material type.

[e]Calculated from spectral data.

REFERENCES TO SECTION 2

I. Definitions of Surface Characteristics

1. Dunkle, R. V., "Thermal Radiation Characteristics of Surfaces", Fundamental Research in Heat Transfer, J. A. Clark, Editor, Macmillan, New York, 1963, pp. 1-31.

2. Edwards, D. K., "Radiative Transfer Characteristics of Materials", J. Heat Transfer, Vol. 91, pp. 1-15, 1969.

3. Nicodemus, F. E., "Directional Reflectance and Emissivity of an Opaque Surface", Applied Optics, Vol. 4, pp. 767-773, 1965.

II. Measurement Techniques

A. Review Papers

4. Dunn, S. T., J. C. Richmond, and J. F. Parmer, "Survey of Infrared Measurement Techniques and Computational Methods in Radiant Heat Transfer", J. Spacecraft and Rockets, Vol. 3, pp. 961-975, 1966.

5. Edwards, D. K., "Thermal Radiation Measurements", Measurements in Heat Transfer , E. R. G. Eckert and R. J. Goldstein, Editors, Hemisphere/ McGraw-Hill, New York, 1976, pp. 425-473.

B. Total Hemispherical Emissivity

6. Nelson, K.E., and J. T. Bevans, "Errors of the Calorimetric Method of Total Emittance Measurement", Measurement of Thermal Radiation Properties of Solids, NASA SP-31, 1963, pp. 55-63.

C. Total Directional Emissivity

7. Snyder, N. W., J. T. Gier, and R. V. Dunkle, "Total Normal Emissivity Measurements on Aircraft Materials Between 100 and 800°F", Trans. Am. Soc. Mech. Engrs., Vol. 77, pp. 1011-1919, 1955.

D. Directional Reflectance

a. Integrating Spheres

8. McNicholas, H. J., "Absolute Methods in Reflectometry", Nat. Bur. Stds. J. Research, Vol. 1, pp. 29-73, 1928.

9. Tingwaldt, C. P., "Über die Messung von Reflexion, Durchlassigkeit und Absorption an Prufkorpern beliebiger Form in der Ulbrichtschen Kugel", Optik, Vol. 9, pp. 323-332, 1952.

10. Toporets, A. S., "Study of Diffuse Reflection from Powders under Diffuse Illumination", Optics and Spectroscopy, Vol. 7, pp. 471-473, 1959.

11. Edwards, D. K., J. T. Gier, K. E. Nelson, and R. D. Roddick, "Integrating Sphere for Imperfectly Diffuse Samples", J. Opt. Soc. Am., Vol. 51, pp. 1279-1288, 1961.

b. Heated Cavities

12. Gier, J. T., R. V. Dunkle, and J. T. Bevans, "Measurement of Absolute Spectral Reflectivity from 1.0 to 15 Microns", J. Opt. Soc. Am., Vol. 44, pp. 558-562, 1954.

13. Tingwaldt, C. P., "Measurement of Spectral Sensitivity in the Infrared", Trans. Conference on the Use of Solar Energy, Vol. III, Part 1, Section A, 1955, pp.57-61.

14. Dunkle, R. V., "Spectral Reflectance Measurements", Surface Effects on Spacecraft Materials, F. J. Clauss, Editor, Wiley, New York, 1960, pp. 117-137.

15. Dunkle, R. V., D. K. Edwards, J. T. Gier, K. E. Nelson, and R. D. Roddick, "Heated Cavity Reflectometer for Angular Reflectance Measurements", Progress in International Research on Thermodynamic and Transport Properties, Am. Soc. Mech. Engrs., New York, 1962, pp. 541-562.

16. Nelson, K. E., E. E. Luedke, and J. T. Bevans, "A Device for the Rapid Measurement of Total Emittance (Reflectance)", J. Spacecraft and Rockets, Vol. 3, pp. 758-760, 1966.

c. Hemispherical Mirrors

17. Birkebak, R. C., J. P. Hartnett, and E. R. G. Eckert, "Measurement of Radiation Properties of Solid Materials", Progress in International Research on Thermodynamic and Transport Properties, Am. Soc. Mech. Engrs. New York, 1962, pp. 563-574.

18. Janssen, J. E., and R. H. Torberg, "Measurement of Spectral Reflectance Using an Integrating Hemisphere", Measurement of Thermal Radiation Properties of Solids, NASA SP-31, 1963, pp. 169-182.

19. White, J. U., "New Method for Measuring Diffuse Reflectance in the Infrared", J. Opt. Soc. Am., Vol. 54, pp. 1332-1337, 1964.

20. Neher, R. T., and D. K. Edwards, "Far Infrared Reflectometer for Imperfectly Diffuse Specimens", Applied Optics, Vol. 4, pp. 775-780, 1965.

21. Edwards, D. K., "Comments on Reciprocal-Mode 2π-Steradian Reflectometers", Applied Optics, Vol. 5, pp. 175-176, 1966.

22. Dunn, S. T., J. C. Richmond, and J. A. Wielbelt, "Ellipsoidal Mirror Reflectometer", J. Opt. Soc. Am., Vol. 55, p. 604, 1965.

23. Wood, B. E., J. G. Pipes, A. M. Smith, and J. A. Roux, "Hemi-Ellipsoidal Mirror Infrared Reflectometer: Development and Operation", Applied Optics, Vol. 15, pp. 940-950, 1976.

E. Bidirectional Reflectance

a. Total Bidirectional Reflectance

24. Schmidt, E., and E. R. G. Eckert, "Über die Richtungsverteilung der Warmestrahlung von Oberflachen", Forshung auf dem Gebiete des Ingenieurwesens, Vol. 6, p. 175, 1935.

25. Eckert, E. R. G., "Messung der Reflexion von Warmestrahlen an Technischen Oberflachen", Forshung auf dem Gebiete des Ingenieurwesens, Vol. 7, pp. 265-270, 1936.

26. Münch, B., "Die Richtungsverteilung bei der Reflexion von Warmestrahlung und ihr Einfluss auf die Warmeubertragung", Mitteilungen aus der Institut fur Thermodynamik und Verbrennungsmotorenbau, No. 16, Verlag Leeman, Zurich, 1955.

b. Spectral Bidirectional Reflectance

27. Birkebak, R. C., and E. R. G. Eckert, "Effects of Roughness of Metal Surfaces on Angular Distribution of Monochromatic Reflected Radiation", J. Heat Transfer, Vol. 87, pp. 85-94, 1965.

28. Torrance, K. E., and E. M. Sparrow, "Off-Specular Peaks in the Directional Distribution of Reflected Thermal Radiation", J. Heat Transfer, Vol. 88, pp. 223-230, 1966.

29. Herold, L. M., and D. K. Edwards, "Bidirectional Characteristics of Rough, Sintered-Metals, and Wire-Screen Surface Systems", AIAA Journal, Vol. 4, pp. 1802-1810, 1966.

30. Birkebak, R. C., "Thermal Radiation Properties of Lunar Materials from the Apollo Missions", Advances in Heat Transfer, Vol. 10, pp. 1-37, 1974.

31. Smith, A. M., "Cryofrost Bidirectional Reflectance", Applications of Cryogenic Technology, Vol. 5, pp. 284-304, 1973.

32. Wood, B. E., and A. M. Smith, "An Investigation of Bidirectional Reflection from Infrared Diffuser Surfaces", Progress in Astronautics and Aeronautics, Vol. 49, pp. 67-84, 1976.

F. Specular Reflectance

33. Bennett, H. E., and W. F. Koehler, "Precision Measurement of Absolute Specular Reflectance with Minimized Systematic Errors", J. Opt. Soc. Am., Vol. 50, pp. 1-6, 1960.

III. Electromagnetic Theory and Polarization

A. Maxwell Equations

34. Stratton, J. A., Electromagnetic Theory, McGraw-Hill, New York, 1941.

B. Useful Approximations

35. Snyder, N. W., "Radiation in Metals", Trans. Am. Soc. Mech. Engrs., Vol. 76, pp. 541-548, 1954.

36. Dunkle, R. V., "Emissivity and Interreflection Relationships for Infinite Parallel Specular Surfaces", Symposium on Thermal Radiation of Solids, NASA SP-55, 1965, pp. 39-44.

37. Seban, R. A., "The Emissivity of Transition Metals in the Infrared", J. Heat Transfer, Vol. 87, pp. 173-176, 1965.

38. Edwards, D. K., and N. Bayard-de-Volo, "Useful Approximations for the Spectral and Total Emissivity of Smooth Bare Metals", Advances in Thermophysical Properties at Extreme Temperatures and Pressures, Am. Soc. Mech. Engrs., New York, 1975, pp. 174-178.

39. Toscano, W. M., and E. G. Cravalho, "Thermal Radiative Properties of the Noble Metals at Cryogenic Temperatures", J. Heat Transfer, Vol. 98, pp. 438-445, 1976.

40. Sievers, A. J., "Thermal Radiation from Metal Surfaces", J. Opt. Soc. Am., Vol. 68, pp. 1505-1518, 1978.

C. Polarization

41. Edwards, D. K., and J. T. Bevans, "Effect of Polarization on Spacecraft Radiation Heat Transfer", AIAA Journal, Vol. 3, pp. 1323-1329, 1965.

42. Edwards, D. K., and R. D. Tobin, "Effect of Polarization on Radiant Heat Transfer through Long Passages", J. Heat Transfer, Vol. 89, pp. 132-138, 1967.

43. Amar, R. C., and D. K. Edwards, "Reflection and Transmission by Rough-Walled Passages", Progress in Astronautics and Aeronautics, Vol. 31, pp. 475-495, 1973.

IV. Thermal Design

44. Camack, W. G., and D. K. Edwards, "Effect of Surface Thermal Radiation Characteristics on the Temperature-Control Problem in Satellites", Surface Effects on Spacecraft Materials, F. J. Clauss, Editor, Wiley, New York, 1960, pp. 3-54.

45. Kreith, F., Radiation Heat Transfer, International Textbook Co., Scranton, Pa., 1962.

46. Edwards, D. K., and R. D. Roddick, "Spectral and Directional Thermal Radiation Characteristics of Surfaces for Heat Rejection by Radiation", Progress in Aeronautics and Astronautics, Vol. 11, pp. 427-446, 1963.

47. Edwards, D. K., and W. M. Hall, "Far Infrared Reflectance of Spacecraft Coatings", Progress in Astronautics and Aeronautics, Vol. 18, pp. 3-19, 1966.

48. Marshall, K. N., and R. A. Breuch, "Optical Solar Reflector: A Highly Stable Low α_S/ε Spacecraft Thermal Control Surface", J. Spacecraft and Rockets, Vol. 5, pp. 1051-1056, 1968.

49. Smith, A. M., "Theory of Radiative Degradation (An Analytical Study of a Solar Degradation Model for Thermal Control Materials and Some Ramifications for Accelerated Solar Radiation Testing)", Progress in Astronautics and Aeronautics, Vol. 21, pp. 639-663, 1969.

50. Kurland, R. M., G. L. Brown, J. F. Thomasson, J. T. Bevans, et al., "Properties of Metallized Flexible Materials in the Space Environment", SAMSO Report TR 78-31, 1978.

51. Edwards, D. K., Solar Collector Design, Franklin Institute Press, Philadelphia, Revised Edition, 1978.

52. Edwards, D. K., A. F. Mills and V. E. Denny, Transfer Processes, Hemisphere/McGraw Hill, New York, Second Edition, 1979.

53. Hottel, H. C., "Radiant Heat Transmission", in W. H. McAdam's Heat Transmission, McGraw-Hill Book Co., 3rd Edition, 1954, Chapter 4 and Table A23.

EXERCISES FOR SECTION 2

1. Suppose that the spectral reflectivity displayed in Fig. 2-8b does not vary with specimen temperature, and suppose that the graph can be extrapolated to infinite wavelength at the value shown last on the curve. Suppose also that the curve for $k/n = 0$ in Fig. 2-4 applies where the spectral reflectivity is low and that for $k/n = 1$ where the spectral reflectivity is high. Calculate exactly the total normal and the total hemispherical emissivity for T = 300 K, 400 K, 600 K, 800 K, and 1000 K. Compare your exact calculations with the "rule" enunciated in the last sentence of the paragraph following Eq. (2-84).

2. The soft-anodized aluminum of Fig. 2-8b is heated to 400 K in a 300 K room. It is viewed by a 300 K radiometer whose signal V is proportional to the difference in radiosity of the surface viewed and σT_R^4 where T_R is the radiometer temperature. The signal V is divided by V_b obtained viewing a black cavity at 400 K, and the ratio V/V_b is reported by the laboratory as the total emissivity of the aluminum. Compare V/V_b with the external total normal emissivity and with the internal total normal emissivity calculated exactly.

3. What is the absorptivity of flame-sprayed alumina at 300 K (Fig. 2-11a) for radiation from

   (a) The extraterrestrial sun.

   (b) The terrestrial sun.

   (c) Radiation from a 2900 K flat tungsten ribbon filament filtered by glass with cut-off wavelengths of 0.33 and 2.7 μm.

   (d) Radiation from a 600 K black body source.

4. Suppose the spectral directional emissivity of a metal were given approximately by the following formula:

$$\varepsilon(\theta,\lambda,T) \doteq 0.5\varepsilon_o \cdot (\lambda_o T/\lambda T_o)^{\frac{1}{2}} \cdot (\cos\theta + \sec\theta)$$

   (a) Find the ratio of hemispherical to normal emissivity.

   (b) Determine how total hemispherical emissivity varies with temperature.

   (c) What assumptions or approximations are necessary to derive the above equation from Eqs. (2-69, 70, and 73)?

   (d) Find the total hemispherical emissivity at 400 K for $\varepsilon_o = 0.05$, $T_o = 300$ K, and $\lambda_o = 10$ μm. Is Eq. (2-79) helpful?

5. A total radiometer has a conical absorber ($\alpha \doteq 1$) exposed to radiation through a 6 mm diameter opening. The detector is conduction-coupled to the radiometer

5. (continued)

case through a silver conductor 10 mm in diameter, 60 μm thick, and 20 mm high. Wire resistors are thermally bonded to the top and bottom of the thermal conductor. These have a resistance of 2200 ohms and change with temperature 4.5×10^{-3} ohms/ohm K. Two load resistors of 2200 ohms are used to complete a bridge circuit with a 10 volt power supply. What signal in millivolts would be detected across the bridge when 1380 W/m^2 are incident upon the 300 K radiometer?

6. A heated cavity reflectometer consists of a 150 mm I.D. cavity 160 mm long. A viewing port 10 by 20 mm is located 90 mm away from a water-cooled sample holder. Data are taken as follows: A smooth reference fin is viewed at 25° off normal by a detector through monochromator optics. The gain of the electronics is adjusted until a reading of $V_r = 1.00$ is obtained. A zero reading V_0 is taken with a shutter closed. The V_r and V_0 readings are checked by opening and closing the shutter and adjusting the gain and zero bias as necessary. Then the specimen is translated to the position formerly occupied by the reference fin. A reading of V_s is recorded, and that reading is taken to be the reflectivity $\rho(\theta=25°,\lambda,T_s)$ of the specimen. The following data are believed to be approximately correct

6. (continued)

Cavity wall temperature: $T_c = 1050$ K

Specimen temperature: $T_s = 305$ K

Reference fin temperature: $T_r = 1045$ K

Shutter and optics box temperature: $T_o = 300$ K

Emissivity of reference fin: $\varepsilon_r = 0.03(\lambda_o T_r/\lambda T_o)^{\frac{1}{2}}$

($\lambda_o = 20$ µm, $T_o = 300$ K)

Answer the following questions for a specimen whose true reflectivity is 0.50:

(a) How much does V_s differ from the true reflectivity ρ due to the presence of the viewing port when the specimen is perfectly diffuse? (Assume $T_r = T_c$ and $T_s = T_o$)

(b) Is there an exit port error when the specimen is perfectly specular?

(c) Suppose the specimen is specular and T_r is equal to T_c. How much error due to sample emission occurs at $\lambda = 2$ µm, 4 µm, 10 µm, 20 µm?

(d) Suppose the specimen is specular and T_s is equal to T_o. How much cold fin error occurs at $\lambda = 2$ µm, 4 µm, 10 µm, and 20 µm?

(e) What is the combined error at $\lambda = 2$ µm, 4 µm, 10 µm, and 20 µm?

7. An inspection solar reflectivity meter consists of a 75 mm diameter integrating sphere with a filtered xenon

7. (continued)

arc-plus-electrode lamp source. While the combination of the filtered arc and integrating sphere efficiency is supposed to approximate the extraterrestrial sun, the simulation is imperfect. Accordingly a filter-wheel of 4 short-wavelength-pass filters is built into the instrument. Filter 1 transmits 90% to 0.45 μm, filter 2, 90% to 0.59 μm, filter 3, 90% to 0.79 μm, and filter 4, 90% to 1.11 μm. The sphere is placed over the specimen; the energy from the lamp is directed onto the sphere wall; and the gain is adjusted to give a reading of 1.00. Then readings V_1, V_2, V_3, and V_4 are taken by rotating filters 1, 2, 3, and 4 successively into the beam. There is found $V_1 = 0.23$, $V_2 = 0.41$, $V_3 = 0.59$, and $V_4 = 0.84$. Then with a filter in the beam, the beam directed on the sphere wall, the gain is increased to give a reading of 100. Then the beam is directed onto the specimen and a reflectivity reading is taken. This procedure is repeated for ρ_1, ρ_2, ρ_3, ρ_4, and (with no filter) ρ_5.

(a) What readings would be expected for the 25 μm thick silicone-alkyd white paint in Fig. 2-6b?

(b) If readings of $\rho_1 = 0.35$, $\rho_2 = 0.50$, $\rho_3 = 0.65$, $\rho_4 = 0.65$, $\rho_5 = 0.63$ are obtained, what is your best estimate of the extraterrestrial solar reflectivity of the specimen?

8. (a) Calculate the hemispherical emissivity into space $n_o = 1$ of a weakly absorbing but thick isothermal dielectric whose index of refraction is $n_1 = 1.5$.

(b) Show that *inside* the dielectric the black-body intensity is n_1^2 times that given by Eq. (1-25) for $n_o = 1$ by equating the emergent net flux on the space side of the interface with incident heat flux on the dielectric side of the interface. (The net flux on the dielectric side is the difference between the incident and reflected fluxes. Incident rays on the dielectric side with angles greater than $\sin^{-1}(n_o/n_1)$ are totally internally reflected.)

9. Estimate the hemispherical absorptivity of silicon carbide from a handbook value of refractive index $n_1 = 2.65$. Assume for the calculation that $k_1 << n_1$.

10. Calculate the total hemispherical emissivity of a metal at 260°C whose optical constants are given by Eq. (2-73) with $r_e = r_{e,o}(T/T_o)$ and $r_{e,o} = 100 \times 10^{-6}$ ohm-cm at 20°C. (Note that Eqs. (2-73) et seq. require the use of absolute temperatures.)

11. A space probe is to be propelled by the reflection of solar photons off a "solar sail" of aluminized plastic film. The plastic film is made $\delta_2 = 2.5$ μm thick of a dielectric with $n_2 = 1.2$, $k_2 = 0$. To keep the sail cool

11. (continued)

during a 0.25 a.u. approach to the sun, the back of the sail is to be given a high emissivity. Accordingly the front side of the plastic is given an optically thick coating of a higher emissivity metal (n_1,k_1) before it is overcoated with aluminum, and the back side is given a thin coating of an absorbing metal with optical constants n_3,k_3. Making use of Kirchhoff's law, find the spectral directional emissivity at $\theta = 0°$ and $\lambda = 6$, 9, 12 and 15 microns when $\delta_3 = 200$ Angstroms, and layers 1 and 3 are the same metal as given in Exercise 10 and at the same temperature of 260°C. (Note that the calculations required are best carried out with a computer having a FORTRAN compiler with complex-number arithmetic.) What thickness δ_3 would give the highest total emissivity at 260°C?

3 RADIATION TRANSFER BETWEEN PERFECTLY DIFFUSE SURFACES

3.A Shape Factors. Recall Eq. (1-22), which should be remembered as being the most important equation in Section 1

$$d^3q^{\pm} = I_\nu^{\pm}\cos\theta d^2\Omega d\nu \tag{3-1}$$

The equation holds for either the radiosity q^+ or the irradiation q^-, hence the ± notation. This fact has an important result for the shape factor; the shape factor has two equally valid and equally important meanings: (1) The shape factor F_{i-j} is the fraction of the radiated power leaving diffuse surface i that is intercepted by surface j. (2) The shape factor F_{i-j} is the fractional weight to be given the radiosity of diffuse surface j in summing up the irradiation upon surface i, or, put another way, it is the fraction of the surrounds of surface i taken up by surface j.

In what immediately follows, the equation will be written without ν or λ subscripts to denote a spectral quantiby, but inspection of Eqs. (1-22 and 23) shows that the same expression holds for either spectral or total quantities.

Consider the first meaning of the shape factor. The power leaving differential surface i is the radiosity times the area

$$d^4Q_i^+ = d^2A_i d^2q_i^+ = d^2A_i I_i^+\cos\theta_i d^2\Omega$$

But for a diffuse surface $I_i^+ = q_i^+/\pi$, Eq. (1-10). The power intercepted by finite surface j is then

$$d^2\dot{Q}_{i-j} = d^2A_i q_i^+ \frac{1}{\pi} \iint_{\Omega_{i-j}} \cos\theta_i d^2\Omega$$

while all the power leaving diffuse surface i is

$$d^2\dot{Q}_{i-j} = d^2A_i q_i^+ \frac{1}{\pi} \iint_{\Omega_{i-j}=0}^{2\pi} \cos\theta_i d^2\Omega = d^2A_i q_i^+$$

Hence the fraction of all the power radiated by diffuse, differential surface i and intercepted by surface j is

$$F_{i-j} = \frac{1}{\pi} \iint_{\Omega_{i-j}} \cos\theta_i d^2\Omega \tag{3-2}$$

The above form of the shape factor definition is particularly convenient when the limits of integration are simple. When the normal to differential surface i coincides with the axis of an axisymmetric surface (such as a disk or sphere), polar-azimuthal angles are used and

$$F_{i-j} = \frac{1}{\pi} \int_0^{2\pi} \int_0^{\theta_{max}} \cos\theta \sin\theta d\theta d\phi = \sin^2\theta_{max} \tag{3-3}$$

When the normal to differential surface i is perpendicular to infinite cylinder j (circular or not), base-axial angles are used and

$$F_{i-j} = \frac{1}{\pi} \int_{\gamma_1}^{\gamma_2} \int_{-\pi/2}^{+\pi/2} \cos^2\beta d\beta \cos\gamma d\gamma = \frac{1}{2} (\sin\gamma_2 - \sin\gamma_1) \tag{3-4}$$

Introduction of the definition of $d^2\Omega$, Eq. (1-4), into (3-2) gives

$$F_{i-j} = \frac{1}{\pi} \iint_{A_j} \frac{\cos\theta_i \cos\theta_j}{R_{i-j}^2} d^2A_j$$

If area A_i is not infinitesimal, one wants the average shape factor for the finite area. Hence

$$F_{i-j} = \frac{1}{A_1} \iint_{A_i} \iint_{A_j} \frac{\cos\theta_i \cos\theta_j}{\pi R_{i-j}^2} d^2A_j d^2A_i \qquad (3\text{-}5)$$

This form of the shape factor definition is general.

An important feature of Eq. (3-5) is shape factor reciprocity

$$A_i F_{i-j} = A_j F_{j-i} \qquad (3\text{-}6)$$

which follows from the symmetry in the double integral. Furthermore it is clear from Eq. (3-2) that

$$\sum_{j=1}^{N} F_{i-j} = 1 \qquad (3\text{-}7)$$

The second meaning of the shape factor can be arrived at by writing Eq. (3-1)--after integration over ν or with ν subscripts--in the form

$$d^2q_i^- = I_i^- \cos\theta_i d^2\Omega$$

But I_i^- is I_j^+ inside the solid angle Ω_{ij} when the fluid is diathermanous. Accordingly

$$d^2q_i^- = q_j^+ \frac{1}{\pi} \cos\theta_i d^2\Omega$$

With the very same definition of the shape factor F_{i-j} appearing in Eq. (3-2) for differential A_i, or in Eq. (3-5) for finite A_i, one can write

$$q_i^- = \sum_{j=1}^{N} q_j^+ F_{i-j} \tag{3-8}$$

Collections of shape factors for elementary geometries have been compiled in a number of works, e.g. [1-5]; a number of devices to enable graphical or photometric evaluation have been described, e.g. [6-8]; and a number of computer programs have been written, e.g. [9,10]. Table 3-1 lists three closed-form solutions that can be used with a programmable calculator.

As an example, consider an inert-gas-filled furnace with floor 4 by 4 meters and ceiling 2 meters high. Number the floor 1, the side walls 2, 3, 4, and 5 and the top 6. Sighting into the furnace with a radiometer or pyrometer allows one to observe directly the radiosities of the walls q_1^+, q_2^+, etc. The radiometer may be calibrated to indicate an apparent temperature $T_{1,app} = (q_1^+/\sigma)^{1/4}$, in which case $q_1^+ = \sigma T_{1,app}^4$. The following results are obtained after many sightings are averaged.

i	$T_{i,app}$ [K]	q_i^+ [W/m^2]
1	500	3544
2	1500	287000
3	1200	117600
4	1200	117600
5	1200	117600
6	1000	56700

TABLE 3-1

CLOSED-FORM EXPRESSIONS OF SELECTED SHAPE FACTORS

Geometry	Expression
1. Opposite rectangles	$F_{1-2} = \frac{2}{\pi XY}\Big\{\ell n \sqrt{\frac{(1+X^2)(1+Y^2)}{1+X^2+Y^2}} + X\sqrt{1+Y^2}\tan^{-1}\frac{X}{\sqrt{1+Y^2}} + Y\sqrt{1+X^2}\tan^{-1}\frac{Y}{\sqrt{1+X^2}} - X\tan^{-1}X - Y\tan^{-1}Y\Big\}$
2. Adjacent rectangles	$F_{1-2} = \frac{1}{4\pi X}\Big\{4X\tan^{-1}\frac{1}{X} + 4Y\tan^{-1}\frac{1}{Y} - 4\sqrt{X^2+Y^2}\tan^{-1}\frac{1}{\sqrt{X^2+Y^2}} + \ell n\frac{(1+X^2)(1+Y^2)}{1+X^2+Y^2} + X^2\ell n\frac{X^2(1+X^2+Y^2)}{(1+X^2)(X^2+Y^2)} + Y^2\ell n\frac{Y^2(1+X^2+Y^2)}{(1+Y^2)(X^2+Y^2)}\Big\}$
3. Parallel, co-axial circles	$F_{1-2} = \frac{1}{2}\left[X - \sqrt{X^2-4Y^2Z^2}\right]$

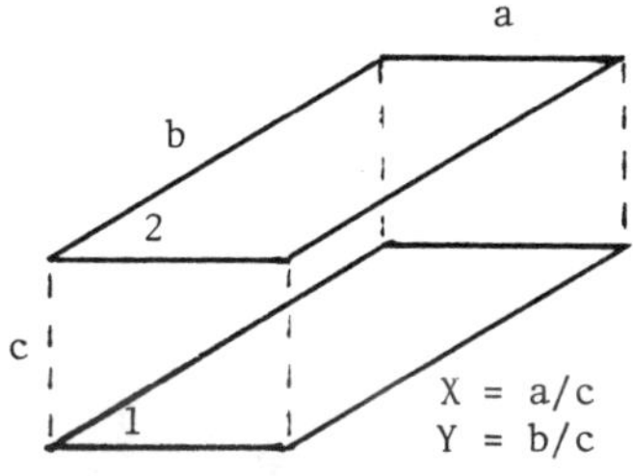

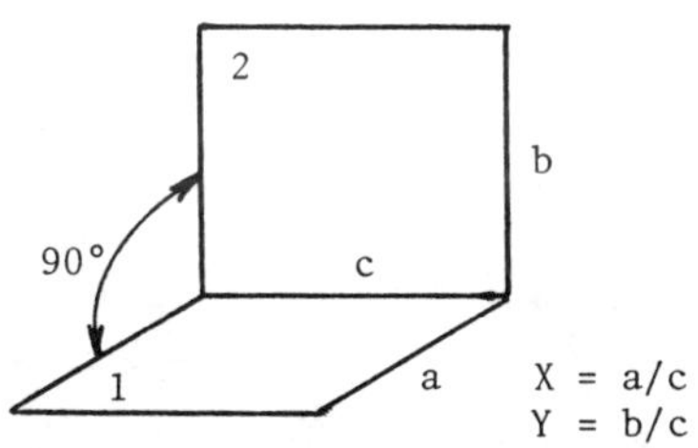

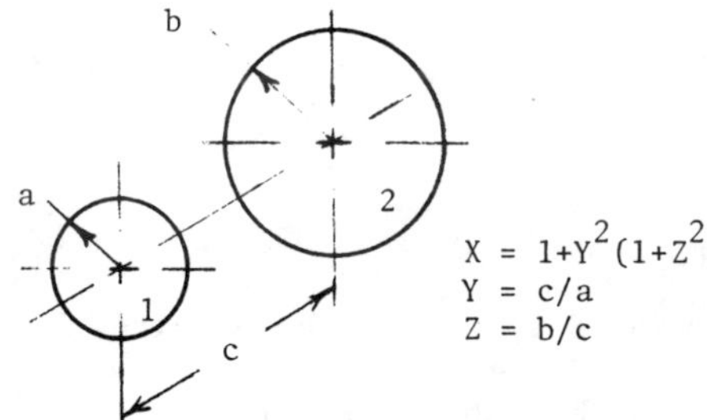

What is the net radiant heat flux into the floor?

The net flux *into* the floor is denoted $-q_1$ to be consistent with the sign convention to be followed later. It is, obviously,

$$-q_1 = q_1^- - q_1^+$$

The quantity q_1^+ is known; the irradiation q_1^- is needed. It is given by Eq. (3-8)

$$q_1^- = q_1^+ F_{1-1} + q_2^+ F_{1-2} + q_3^+ F_{1-3} + q_4^+ F_{1-4} + q_5^+ F_{1-5} + q_6^+ F_{1-6}$$

All that is needed are the shape factors F_{i-j}. Because no solid-angle-cosine product is subtended by surface 1 itself from any point upon surface 1, F_{1-1} is clearly zero; the surface cannot "see" itself. Table 3-1 gives one relation for F_{i-j} where j = 2,3,4, or 5, and another for F_{1-6}. For the former

$$F_{1-2} = \frac{1}{4\pi X}\left\{4X\tan^{-1}\left(\frac{1}{X}\right) + \ldots\right\} = 0.14619$$

where X = 4/4 = 1 and Y = 2/4 = 1/2. For the latter

$$F_{1-6} = \frac{2}{\pi XY}\left\{\log\left[\frac{(1+X^2)(1+Y^2)}{1+X^2+Y^2}\right]^{1/2} + \ldots\right\} = 0.41525$$

where X = 4/2 = 2 and Y = 4/2 = 2. As a check we note that Eq. (3-7) is satisfied.

$$F_{1-1} + F_{1-2} + \ldots F_{1-6} = 4(0.14619) + 0.41525 = 1.00$$

It was not necessary in view of Eq. (3-7) to have calculated F_{1-2} from the formula in the table; from the value of

F_{1-6}; the equation, and the symmetry in the problem, the result could have been obtained; but the check helps guard against errors. To complete the example

$$q_1^- = (0.14619)(287000)+(3)(0.14619)(117600)$$

$$+ (0.41525)(56700) = 117070 \text{ W/m}^2$$

$$-q_1 = 117070-3540 = 113500 \text{ W/m}^2$$

$$-\dot{Q}_1 = -q_1A_1 = (113500)(16) = 1816 \text{ kW}$$

$$(6.2\text{x}10^6 \text{ Btu/hr})$$

The example shows the utility of the shape factor as a weighting factor in Eq. (3-8) and illustrates the use of Table 3-1 and Eq. (3-7).

The value of the relations shown in Table 3-1 is greatly increased by consideration of shape factor algebra. Denote A_iF_{i-j} in Eqs. (3-5 and 6) as

$$G_{i-j} = G_{j-i} \equiv A_iF_{i-j} = \iint_{A_i} \iint_{A_j} \frac{\cos\theta_i \cos\theta_j}{\pi R_{i-j}^2} d^2A_j d^2A_i \tag{3-9}$$

Suppose area A_j consists of two subareas A_k and A_ℓ. Let the subscript $(k+\ell)$ denote the total surface j, and let K_{i-j} denote the shape factor kernel

$$K_{i-j} = \frac{\cos\theta_i \cos\theta_j}{\pi R_{i-j}^2}$$

Then

$$G_{i-(k+\ell)} = \iint_{A_i} \left\{ \iint_{A_k} K_{i-k} d^2A_k + \iint_{A_\ell} K_{i-\ell} d^2A_\ell \right\} d^2A_i$$

$$G_{i-(k+\ell)} = G_{i-k} + G_{i-\ell} \tag{3-10}$$

This relation makes it possible to find A_1F_{1-3} illustrated in Fig. 3-1. A programmable calculator having a routine to find G_{i-k} can easily be used to find $G_{i-(k+\ell)}$ and G_{i-k} and then by subtraction $G_{i-\ell}$.

Figure 3-2 shows a somewhat more involved situation. One must recognize from symmetry that $G_{1-3} = G_{4-2}$; then the rest is easy. The fact that G_{1-3} equals G_{4-2} is obvious for the opposed rectangles shown. That G_{1-3} equals G_{4-2} for adjacent unequal rectangles is not so obvious. However, one way to proceed to show the relation is to divide the rectangles into n equal strips, numbered from left to right, so that the right-hand edge of the mth strip nearly coincides with the line dividing large area 1 from large area 2. In the limit as n becomes large the coincidence becomes exact. Starting with an adjacent pair of equal strips and successively adding others allows one to prove a lemma that $G_{1-[2'+3'+\ldots+k']} = G_{[2+3+\ldots+k]-1'}$ where the unprimed numbers here refer to the small strips in one plane and the primed ones to the other. With the aid of this lemma the theorem $G_{[1+2+\ldots+m]-[(m+1)'+\ldots+n']} = G_{[(m+1)+\ldots+n]-[1'+]'+\ldots+m']}$ is proved by adding $G_{1-[2'+\ldots+n']}$ to $G_{2-[3'+\ldots+n']}$, cancelling the $G_{1-2'}$ and $G_{2'-1}$ terms and continuing the addition until the mth term is included.

Figure 3-3 shows two cases of more general interest.

FIGURE 3-1

Simple Cases of Shape Factor Algebra

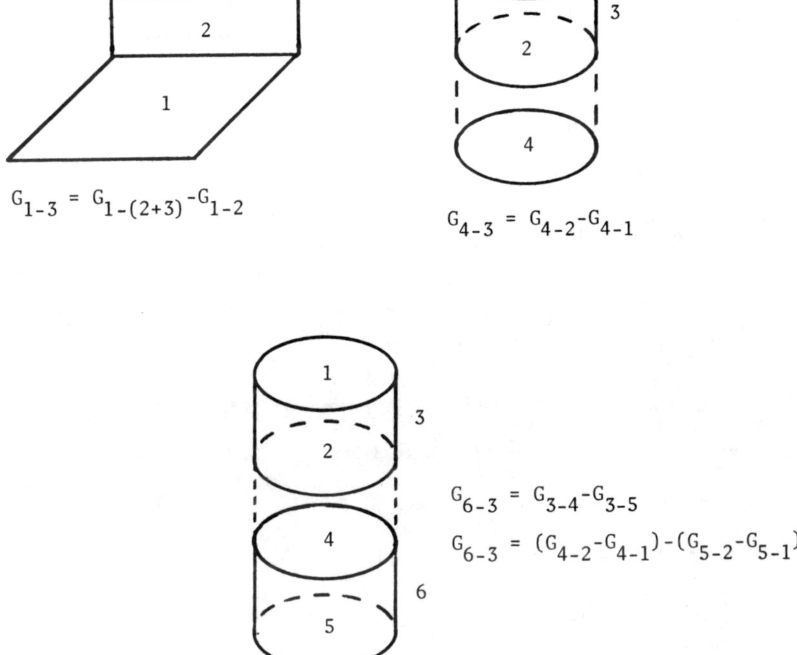

FIGURE 3-2

Shape Factor Algebra for Offset Rectangles with Shared Z-Limit

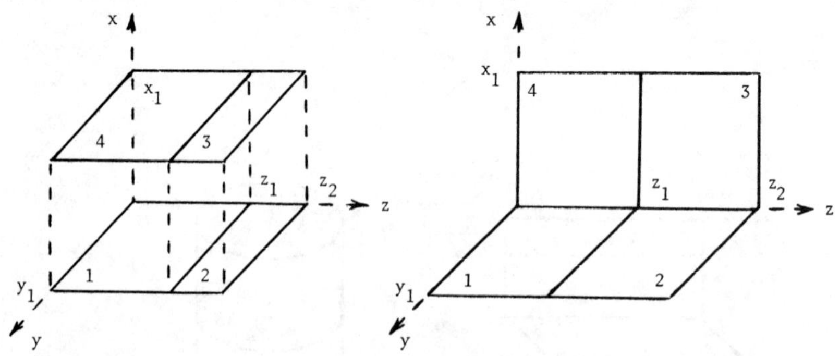

Opposite Planes Adjacent Planes

Area-Number Notation:

$$G_{1-3} = \frac{1}{2}\,[G_{(1+2)-(3+4)} - G_{1-4} - G_{2-3}]$$

x-y-z-Limit Notation:

$$G^{(2)}(x_1,y_1,z_1,z_2) = \frac{1}{2}\,[G^{(1)}(x_1,y_1,z_2) - G^{(1)}(x_1,y_1,z_1) - G^{(1)}(x_1,y_1,z_2-z_1)]$$

where for opposite rectangles, Table 3-1, item 1

$$G^{(1)}(c,a,b) = abF_{1-2}(X,Y)\ ,\quad X = a/c\ ,\quad Y = b/c$$

and for adjacent rectangles, Table 3-1, item 2

$$G^{(1)}(b,a,c) = acF_{1-2}(X,Y)\ ,\quad X = a/c\ ,\quad Y = b/c$$

FIGURE 3-3

Shape Factor Algebra for Rectangles in Opposite and Adjacent Planes

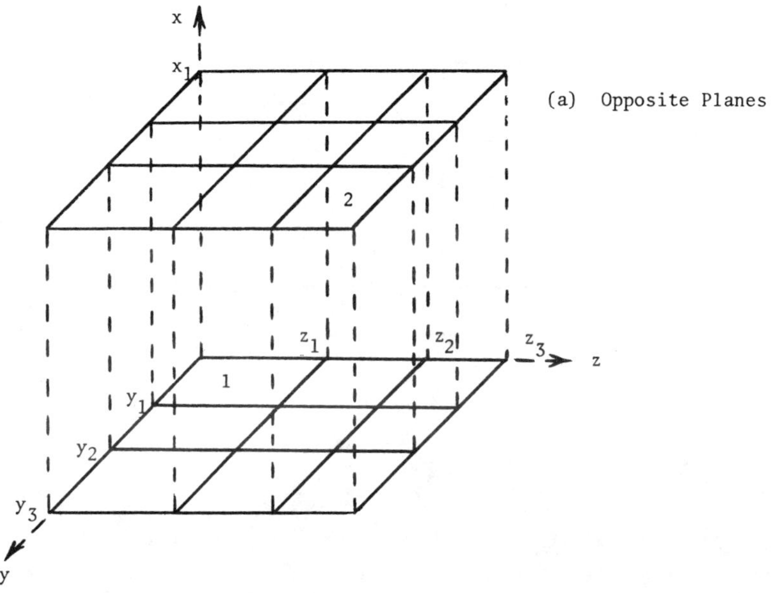

(a) Opposite Planes

$$G_{1-2} = G^{(5)}(x_1,y_1,y_2,y_3,z_1,z_2,z_3) = G^{(4)}(x_1,y_1,y_3,z_1,z_2,z_3) - G^{(4)}(x_1,y_1,y_2,z_1,z_2,z_3)$$

$$G^{(4)}(x_1,y_1,y_2,z_1,z_2,z_3) = G^{(3)}(x_1,y_1,y_2,z_1,z_3) - G^{(3)}(x_1,y_1,y_2,z_1,z_2)$$

$$G^{(3)}(x_1,y_1,y_2,z_1,z_2) = \frac{1}{2}\,[G^{(2)}(x_1,y_2,z_1,z_2) - G^{(2)}(x_1,y_1,z_1,z_2) - G^{(2)}(x_1,y_2-y_1,z_1,z_2)]$$

$G^{(2)}(x_1,y_1,z_1,z_2)$ = See Figure 3-2

FIGURE 3-3 (continued)

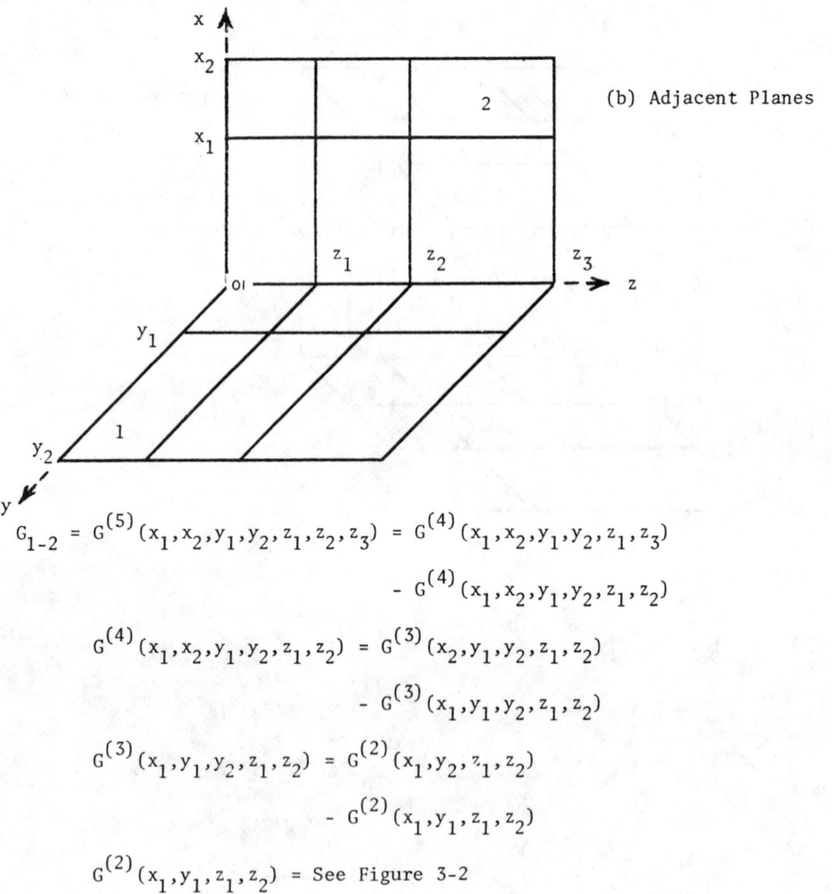

$$G_{1-2} = G^{(5)}(x_1,x_2,y_1,y_2,z_1,z_2,z_3) = G^{(4)}(x_1,x_2,y_1,y_2,z_1,z_3) - G^{(4)}(x_1,x_2,y_1,y_2,z_1,z_2)$$

$$G^{(4)}(x_1,x_2,y_1,y_2,z_1,z_2) = G^{(3)}(x_2,y_1,y_2,z_1,z_2) - G^{(3)}(x_1,y_1,y_2,z_1,z_2)$$

$$G^{(3)}(x_1,y_1,y_2,z_1,z_2) = G^{(2)}(x_1,y_2,z_1,z_2) - G^{(2)}(x_1,y_1,z_1,z_2)$$

$G^{(2)}(x_1,y_1,z_1,z_2)$ = See Figure 3-2

Shape factors between long parallel surfaces such as nuclear fuel rods or boiler tubes are easily obtained via Hottel's String Rule [3]. The rule is that the shape factor from (circular or noncircular) cylinder 1 of length $L \to \infty$ and perimeter P_1 to parallel (circular or noncircular) cylinder 2 of length $L \to \infty$ is the sum of the lengths of diagonal strings D_1 and D_2 minus the lateral strings L_1 and L_2 divided by $2P_1$.

$$F_{1-2} = \frac{D_1 + D_2 - L_1 - L_2}{2P_1} \tag{3-11}$$

The diagonals and laterals are measured in the plane of the cross-section as shown in Fig. 3-4. The diagonals are the crossed strings between the extremities of the two surfaces and the laterals are the uncrossed ones. The strings are understood to be pulled tight and may lie along the surface of the body as necessary. For example if surface 1 is the *whole* cylinder ABCDEA' and surface 2 the *half* cylinder HIJF, then the extremities of 1 are understood to be A and A' and those of surface 2 are F and H. The diagonals are ABCIH and A'EDJF, and the laterals ABF and A'EH. The perimeter P_1 is ABCDEA', and P_2 is HIJF.

FIGURE 3-4

The String Rule Diagonals and Laterals

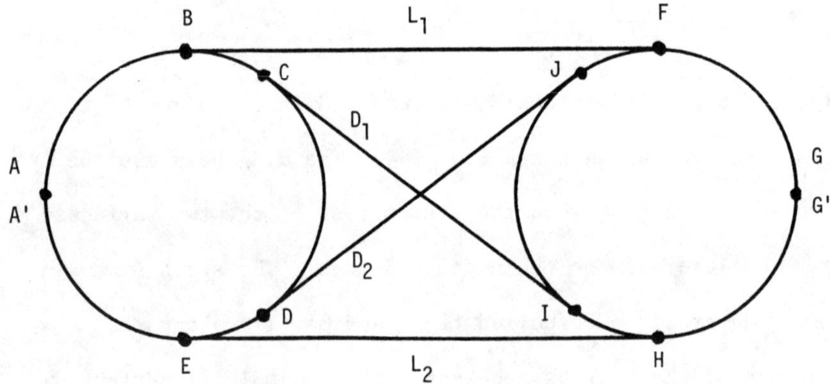

3.B Radiosity-Irradiation Formulations. The example in Section 3.A illustrating the use of shape factors took values of radiosity to be known from direct measurements. If it is not possible or convenient to make such measurements, or if one wants a model to investigate the results of design changes, it is possible to calculate the radiosity, provided that the wall temperature or heat flux (or a relation between them) is known. In what follows the ν or λ subscripts continue to be left off for convenience, with the understanding that the equations hold not only for spectral values in general, but also for total values when the radiation characteristics are approximated as gray. The surfaces are assumed to be perfectly diffuse and opaque.

The engineering radiation transfer problem is formulated for an enclosure of n discrete surfaces; if necessary an imaginary black surface at ambient temperature is used to complete the enclosure, when the surrounds outside the otherwise incomplete enclosure constitute a hohlraum at the ambient temperature. In some cases (notably the hollow sphere, the hollow cylinder, or infinite parallel plates) a useful exact formulation is achieved by passing from the discrete (and therefore approximate) surface representation with sums of finite terms to a continuous representation with integrals of differential terms. One could commence with the exact integral formulation and pass to the finite one. To be brief we commence with and confine ourselves to the discrete surface case.

The relations necessary to formulate the enclosure problem are few in number. First is simply the net outward flux relation Eq. (1-3), here rewritten for the ith surface

$$q_i = q_i^+ - q_i^- \tag{3-12}$$

Second is the definition of radiosity for a diffuse opaque surface, the sum of emitted plus reflected radiation,

$$q_i^+ = \varepsilon_i B_i + \rho_i q_i^- \tag{3-13}$$

where B_i is used to stand for $q_b^+(T_i) = \pi I_b^+(T_i)$ to simplify notation. Finally there is the introduction of the shape factor through Eq. (3-9)

$$q_i^- = \sum_{j=1}^{n} F_{i-j} q_j^+ \tag{3-14}$$

Of course, with these three equations one has a few auxiliary relations. For the near-equilibrium-state, opaque surface there holds

$$\rho_i = 1 - \alpha_i = 1 - \varepsilon_i \tag{3-15}$$

The shape factor relations also are pertinent

$$A_i F_{i-j} = A_j F_{j-i} \tag{3-16}$$

$$\sum_{j=1}^{n} F_{i-j} = 1 \tag{3-17}$$

While only three equations, Eqs. (3-12 to 14), describe the enclosure problem, there are more that can be derived from them. Chief among these are those relating the heat flux to the other quantities. From Eqs. (3-12, 13, and 15) one has

$$q_i = \varepsilon_i(B_i - q_i^-) \tag{3-18}$$

$$q_i^- = B_i - q_i/\varepsilon_i \tag{3-19}$$

$$q_i = [\varepsilon_i/\rho_i][B_i - q_i^+] \tag{3-20}$$

$$q_i^+ = B_i - [\rho_i/\varepsilon_i]q_i \tag{3-21}$$

From Eqs. (3-12, 14, and 17) one has

$$q_i = q_i^+ - \sum_{j=1}^{n} F_{i-j}q_j^+ = \sum_{j=1}^{n} F_{i-j}(q_i^+ - q_j^+) \tag{3-22}$$

Returning to the original Eqs. (3-12 to 14), we see that there are n values of i for each of the three equations, 3n equations in all. There are 3n unknowns, n values each of the heat flux q_i, radiosity q_i^+, and irradiation q_i^-. Inspection of the equations shows that only one, Eq. (3-14), involves n terms; the others relate to only the three quantities for each i. Hence one may easily reduce the 3n equations in 3n unknowns to a set of n equations in n unknowns. One has the choice of formulating in q_i^-, q_i^+, or conceivably, some linear combination of them.

If one formulates in q_i^-, one substitutes Eq. (3-13) into the right-hand side of Eq. (3-14) to obtain

$$\sum_{j=1}^{n} [\delta_{i,j} - \rho_j F_{i-j}]q_j^- = \sum_{j=1}^{n} F_{i-j}\varepsilon_j B_j \tag{3-23}$$

where $\delta_{i,j}$ is the delta function

$$\delta_{i,j} = \begin{cases} 1 & j=i \\ 0 & j \neq i \end{cases} \tag{3-24}$$

If one formulates in q_i^+ one substitutes Eq. (3-13) into the left-hand side of Eq. (3-14) to obtain

$$\sum_{j=1}^{n} [\delta_{i,j} - \rho_i F_{i-j}] q_j^+ = \varepsilon_i B_i \qquad (3\text{-}25)$$

If one formulates in q_i, one substitutes Eqs. (3-19 and 20) into Eq. (3-14) to obtain

$$B_i - q_i/\varepsilon_i = \sum_{j=1}^{n} F_{i-j} [B_j - (\rho_j/\varepsilon_j) q_j]$$

Multiplying by A_i, rearranging, and taking the heat flows $\dot{Q}_i = A_i q_i$ instead of the heat fluxes q_i as the unknown quantities gives

$$\sum_{j=1}^{n} [(\delta_{i,j}/\varepsilon_i) - (\rho_j/\varepsilon_j A_j) A_i F_{i-j}] \dot{Q}_j$$

$$= \sum_{j=1}^{n} A_i F_{i-j} (B_i - B_j) \qquad (3\text{-}26)$$

How one proceeds with the mechanics of obtaining a solution depends upon one's goals. If one wants merely a set of heat flows for one particular problem, one might simply use Eqs. (3-13 and 14) as they stand in an iterative numerical calculation. One might begin by assuming the irradiation is zero (or that it is $\Sigma\varepsilon_j A_j B_j / \Sigma\varepsilon_j A_j$ from the exact solution for surfaces on the inside of a sphere), then calculate a set of radiosities q_i^+ from Eq. (3-13), then calculate a set of irradiations q_i^- from Eq. (3-14), and then repeat until the solutions converge. Then Eq. (3-18 or

20) can be used to find the heat fluxes q_i, and those multiplied by area to give $\dot{Q}_i$.

More often, one wants not merely a set of heat flows for one particular set of temperatures, but a set of parameters that describe the system behavior for any set of temperatures. The system parameters are the transfer factors $\mathscr{F}_{i-j}$ where

$$\dot{Q}_i = \sum_{j=1}^{n} A_i \mathscr{F}_{i-j} [B_i - B_j] \tag{3-27}$$

An important property of the transfer factor is

$$A_i \mathscr{F}_{i-j} = A_j \mathscr{F}_{j-i} \tag{3-28}$$

That the heat flows are linearly related to the blackbody radiosity differences, as indicated by Eq. (3-27), derives from the linearity of the governing equations. The linearity permits one, if it is convenient, to use the powerful superposition principle. One may think of $A_i \mathscr{F}_{i-j}$ as the heat flow into surface j (i.e. $-\dot{Q}_j$) when B_i is set equal to unity and all other B_j are set equal to zero. This concept, for example, shows that one could evaluate $A_i \mathscr{F}_{i-j}$ numerically via the iterative procedure described after Eq. (3-26). It follows from this concept that

$$\sum_{j=1}^{n} A_i \mathscr{F}_{i-j} = \varepsilon_i A_i \tag{3-29}$$

Comparison of Eq. (3-26) with $\rho_i = 0$ and $\varepsilon_i = 1$ with Eq. (3-27) shows that the transfer factor and shape factor are identical when all the walls are black. When all the walls are black, radiation is emitted by a hot surface i at the full black-body rate, and only that radiation emitted directly toward surface j is transferred to it. When the walls are not black the emitted flux at i is only $\varepsilon_i B_i$ and surface j does not absorb all of the irradiation received upon it, but radiation from i is reflected by other surfaces, and some of this radiation impinges upon j. In general, the transfer factor $\mathscr{F}_{i-j}$ is increased by increasing ε_i and ε_j (which one can regard as the emissivity and absorptivity respectively) and by lowering ε_k (hence increasing ρ_k) of all other surfaces $k \neq i$ or j.

Matrix inversion and multiplication with the aid of a high-speed, large-storage-capacity computer is the means usually employed to obtain the set of transfer factors. Any of the three equations, Eqs. (3-23, 25, or 26), may be used. For example, Eq. (3-23) can be written in matrix form

$$|m_{i,j}||q_j^-| = |F_{i,j}||b_j| \tag{3-30}$$

where $|m_{i,j}|$ denotes a matrix

$$|m_{i,j}| = \begin{vmatrix} m_{1,1} & m_{1,2} & \cdots & m_{1,n} \\ m_{2,1} & m_{2,2} & \cdots & m_{2,n} \\ \cdot & & & \\ \cdot & & & \\ \cdot & & & \\ m_{n,1} & m_{n,2} & \cdots & m_{n,n} \end{vmatrix} \tag{3-31}$$

and an element is, according to Eq. (3-23)

$$m_{i,j} = \delta_{i,j} - \rho_i F_{i-j} \tag{3-32}$$

An element of matrix $|F_{i,j}|$ is the shape factor

$$F_{i,j} = F_{i-j} \tag{3-33}$$

Matrices $|q_j^-|$ and $|b_j|$ are column matrices

$$|q_j^-| = \begin{vmatrix} q_1^- \\ q_2^- \\ \cdot \\ \cdot \\ \cdot \\ q_n^- \end{vmatrix} \tag{3-34}$$

$$|b_j| \equiv \begin{vmatrix} \varepsilon_1 B_1 \\ \varepsilon_2 B_2 \\ \cdot \\ \cdot \\ \cdot \\ \varepsilon_n B_n \end{vmatrix} \tag{3-35}$$

The rule for matrix multiplication is as follows: If matrix $|c_{i,j}|$ is the product of two matrices $|a_{i,k}|$ and $|b_{k,j}|$

$$|c_{i,j}| = |a_{i,k}||b_{k,j}| \tag{3-36}$$

then an element of matrix $|c_{i,j}|$ is

$$c_{i,j} = \sum_{k=1}^{n} a_{i,k} b_{k,j} \tag{3-37}$$

Standard computer routines exist for obtaining $|M_{i,j}|$, the inverse of $|m_{i,j}|$, so that one can obtain

$$|q_i^-| = |M_{i,k}||F_{k,j}| \cdot |b_j| \tag{3-38}$$

Now the superposition principle is useful. If the only value of B_j not equal to zero is $B_k = 1$, then

$$q_i^- = M_{i,1}F_{1-k}\varepsilon_k + M_{i,2}F_{2-k}\varepsilon_k + \ldots + M_{i,n}F_{n-k}\varepsilon_k$$

But we understand $\varepsilon_i q_i^-$ to be $\mathscr{F}_{i-k}$ under this circumstance. Accordingly

$$\mathscr{F}_{i-k} = \varepsilon_i\varepsilon_k[M_{i,1}F_{1,k}+M_{i,2}F_{2,k}+\ldots+M_{i,n}F_{n,k}] \tag{3-39}$$

We see that each element of the transfer factor matrix $|\mathscr{F}_{i,k}|$ is found by taking the corresponding element from the matrix product

$$|P_{i,k}| = |M_{i,j}||F_{j,k}| \tag{3-40}$$

and multiplying it by $\varepsilon_i\varepsilon_k$,

$$\mathscr{F}_{i-k} = \varepsilon_i\varepsilon_k P_{i,k} \tag{3-41}$$

Alternatively one may invert the matrix defined by Eq. (3-25). Equation (3-20) then indicates that the transfer factors are inverse matrix elements multiplied by $\varepsilon_i\varepsilon_k/\rho_i$. While this result is more directly obtained, it is not useful for $\rho_i = 0$. Instead of using Eq. (3-20) one could use Eq. (3-22) to avoid the difficulty posed by $\rho_i = 0$, but at the expense of somewhat more complexity.

The matrix inversion technique is needed only when n is large, say four or more. For fewer surfaces, a simpler means of solution, and one which conveys a great deal more physical insight to aid the designer in exercising creativity, is the network method described in the next section. Nevertheless, a matrix example with n = 2 is presented. Consider a cubical enclosure with one face being surface 1 at a known temperature and with known emissivity ε_1 and the other five faces together being surface 2 with emissivity ε_2. Suppose $\varepsilon_1 = \varepsilon_2 = 0.5$, and $A_1 = 1\ m^2$. To two decimal places $F_{2-2} = 0.80$, and $F_{2-1} = 0.20$. The matrix $|m_{i,j}|$ is, from Eq. (3-32),

$$|m_{i,j}| = \begin{vmatrix} +1.00 & -0.50 \\ -0.10 & +0.60 \end{vmatrix}$$

The inverse matrix is

$$|M_{i,j}| = \begin{vmatrix} +\dfrac{0.60}{0.55} & +\dfrac{0.50}{0.55} \\[2ex] +\dfrac{0.10}{0.55} & +\dfrac{1.00}{0.55} \end{vmatrix}$$

The matrix $|P_{i,k}|$ is, from Eq. (3-40)

$$|P_{i,k}| = \begin{vmatrix} \dfrac{12}{11} & \dfrac{10}{11} \\[2ex] \dfrac{2}{11} & \dfrac{20}{11} \end{vmatrix} \cdot \begin{vmatrix} 0 & 1 \\[2ex] \dfrac{1}{5} & \dfrac{4}{5} \end{vmatrix} = \begin{vmatrix} \dfrac{2}{11} & \dfrac{20}{11} \\[2ex] \dfrac{4}{11} & \dfrac{18}{11} \end{vmatrix}$$

The $|\mathscr{F}_{i-k}|$ matrix is, from Eq. (3-41)

$$|\mathscr{F}_{i-k}| = \begin{vmatrix} \dfrac{1}{22} & \dfrac{5}{11} \\[2ex] \dfrac{1}{11} & \dfrac{9}{22} \end{vmatrix}$$

As a check we observe that $A_1 \mathscr{F}_{1-2} = 5/11 = A_2 \mathscr{F}_{2-1} = 5/11$, satisfying Eq. (3-28). As a further check Eq. (3-29) is verified.

$$A_1 \mathscr{F}_{1-1} + A_1 \mathscr{F}_{1-2} = \frac{1}{22} + \frac{5}{11} = \frac{1}{2} = \varepsilon_1 A_1$$

$$A_2\mathscr{F}_{2-1} + A_2\mathscr{F}_{2-2} = \frac{5}{11} + \frac{45}{22} = \frac{5}{2} = \varepsilon_2 A_2$$

If the surfaces are gray and T_1 = 1500 K (2700°F) and T_2 = 1000 K (1800°R), then from Eq. (3-27)

$$\dot{Q}_1 = A_1\mathscr{F}_{1-2}(\sigma T_1{}^4 - \sigma T_2{}^4)$$

$$\dot{Q}_1 = (5/11)(5.67\text{x}10^{-8})(1500^4 - 1000^4)$$

$$\dot{Q}_1 = 104.7 \text{ kW } (357000 \text{ Btu/hr})$$

3.C Refractory Surfaces. A perfectly insulated or adiabatic surface is said to be a refractory. For such a surface

$$q_i = 0 \tag{3-42}$$

and hence from Eq. (3-12)

$$q_i^+ = q_i^- \tag{3-43}$$

and from Eqs. (3-12), 13, and 15)

$$B_i = q_i^- = q_i^+ \tag{3-44}$$

Note that these relations hold regardless of the value of surface emissivity. The refractory surface can be equally well thought of as a perfect reflector or a perfect absorber and reradiator. This statement is made, of course, subject to the assumptions of perfectly diffuse surfaces. Furthermore, if the surface is nongray, it is not adiabatic on a spectral basis even though it may be on a total basis.

If surface k is a refractory surface, the kth row of matrix $|m_{i,j}|$ has elements according to Eqs. (3-14 and 43)

$$m_{k,j} = \delta_{k,j} - F_{k-j} \tag{3-45}$$

and

$$b_j = 0 \tag{3-46}$$

These changes can be incorporated into Eqs. (3-38 and 41). However, Eq. (3-41) applies only to the nonrefractory surfaces; for a refractory surface i = k, $\mathcal{F}_{k-j}$ is zero.

If surfaces i and j are not refractories, then, if surface k is made a refractory, the transfer factor $\mathcal{F}_{i-j}$ is increased, because the refractory acts as a perfect reflector. As discussed previously a reflecting surface increases the transfer factor between the other surfaces present in the enclosure.

3.D <u>The Radiation Network</u>. The facts that heat is conserved (just like electrical charge) and that radiant heat flows are proportional to differences in radiosity (just like electrical current being proportional to voltage differences) mean that an electrical analog to the radiation heat transfer problem can be constructed [17,18]. Heat is conserved at each of the n surfaces. These n surfaces are regarded as n nodes. The heat flows at each node sum to zero. The heat flow *in* across the m-control-surface of the ith node by conduction is $\dot{Q}_i$. This heat flow must equal the heat flow *out* across the s-control surface.

Equation (3-22) states that the heat flow out across the s-surface is

$$\dot{Q}_i = q_i A_i = \sum_{j=1}^{n} A_i F_{i-j}(q_i^+ - q_j^+) \tag{3-47}$$

It is apparent from this equation that radiosity q_i^+ is to be considered the voltage or potential at the ith node, and this potential drives heat flows into the other nodes at potentials q_j^+ through internodal conductances A_iF_{i-j}, that is, through internodal resistances

$$R_{i-j} = \frac{1}{A_iF_{i-j}} \tag{3-48}$$

The relation between surface heat flux and radiosity is given by Eq. (3-20). Multiplying by area A_i to obtain heat flow gives

$$\dot{Q}_i = [\varepsilon_iA_i/\rho_i][B_i-q_i^+] \tag{3-49}$$

This equation tells us to regard the heat flow into node i as driven by black-body radiosity B_i through surface resistance R_i

$$R_i = \frac{\rho_i}{\varepsilon_iA_i} = \frac{1-\varepsilon_i}{\varepsilon_iA_i} \tag{3-50}$$

and against the node potential q_i^+.

Figure 3-5 shows the circuit connections for a single node i. To construct a radiation network for an enclosure, one places the requisite number (n) of nodes on a diagram and draws all interconnections of the R_{i-j} form. One then draws the surface resistances of the form R_i and the requisite potentials B_i. Figure 3-6 shows a network for n = 3.

If n = 2, or if n = 3 but the surface represented by node i = 3 is a refractory so that it is open circuit, the network is readily reduced to a single equivalent resistance

FIGURE 3-5

Radiation Network Connections for Node i

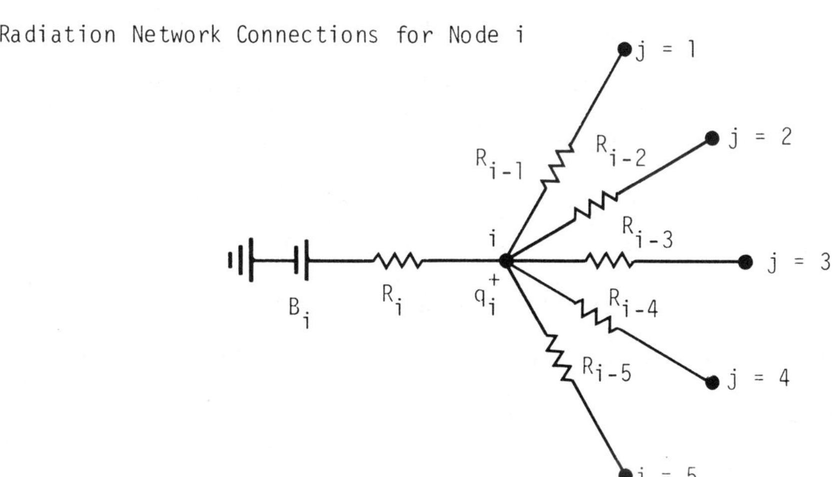

FIGURE 3-6

The Radiation Network for n = 3

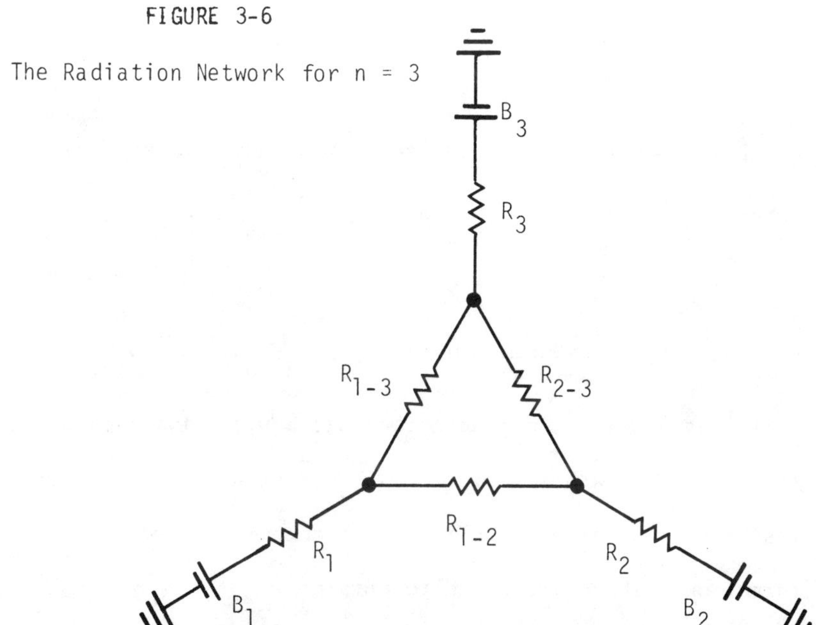

between potentials B_1 and B_2. That equivalent resistance is $[1/A_1\mathscr{F}_{1-2}]$. The meaning of the transfer factor is very clear in this context. In reducing the simple series-parallel network, the rules simply are two: (1) Resistances in series add. (2) Conductances in parallel add (conductance is the reciprocal of resistance).

For n = 3 the delta-wye and wye-delta transformations may be used to find the transfer factors. Figure 3-7 shows the transformations.

For example consider a cubical enclosure with surface 1 the floor at 500 K (900°F), surface 2 a side at 1500 K (2700°F), and surface 3 the remaining three sides and top at 1000 K (1800°F). All surfaces are gray with an emissivity of 0.50. Area A_1 is taken to be 1 m^2, and to two decimal places $F_{1-2} = 0.2$, $F_{1-3} = 0.8$, $F_{2-3} = 0.8$. Figure 3-2 shows the radiation network. There is found

$$R_{1-2} = \frac{1}{(1)(0.2)} = 5.0 \ , \quad R_{1-3} = R_{2-3} \frac{1}{(1)(0.8)} = 1.25$$

$$R_1 = R_2 = \frac{1-0.5}{(1)(0.5)} = 1.0 \ , \ R_3 = \frac{1-0.5}{(4)(0.5)} = 0.25$$

Following a delta-wye transformation, a wye network is found to exist with each leg of the wye consisting of the surface resistance in series with the resistance found from the transformation. These are added to obtain

$$R_{1,tot} = R_{2,tot} = 1.00 + 0.8\bar{3} = 1.8\bar{3}$$

$$R_{3,tot} = 0.25 + 0.208\bar{3} = 0.458\bar{3}$$

FIGURE 3-7

The Delta-Wye and Wye-Delta Transformations

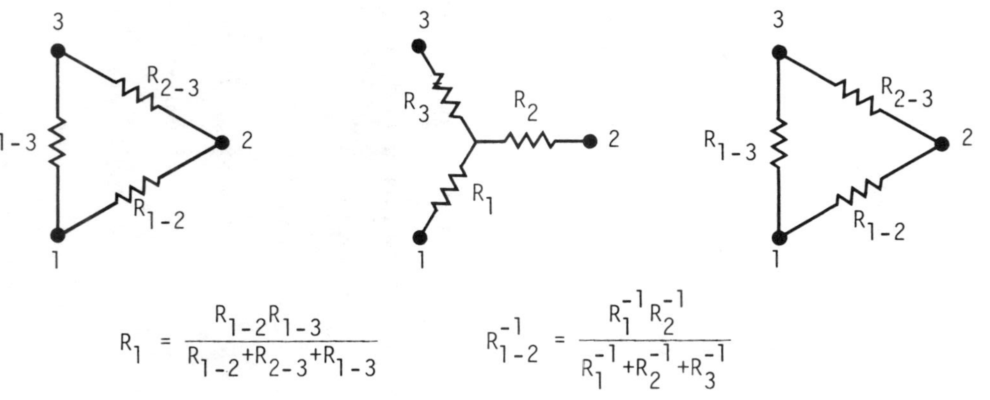

$$R_1 = \frac{R_{1-2}R_{1-3}}{R_{1-2}+R_{2-3}+R_{1-3}} \qquad R^{-1}_{1-2} = \frac{R_1^{-1}R_2^{-1}}{R_1^{-1}+R_2^{-1}+R_3^{-1}}$$

Then a wye-delta transformation is made to obtain the transfer factors (in units of m^2)

$$A_1\mathscr{F}_{1-2} = 0.0909 \ , \ A_1\mathscr{F}_{1-3} = 0.3636 \ , \ A_2\mathscr{F}_{2-3} = 0.3636$$

The heat flows are then found with Eq. (3-47)

$$\dot{Q}_1 = A_1\mathscr{F}_{1-2}(B_1-B_2) + A_1\mathscr{F}_{1-3}(B_1-B_3)$$

$$\dot{Q}_1 = 0.0909(3540-287000) + 0.3636(3540-56700)$$

$$\dot{Q}_1 = -45.1 \text{ kW}$$

$$\dot{Q}_2 = A_2\mathscr{F}_{2-1}(B_2-B_1) + A_2\mathscr{F}_{2-3}(B_2-B_3)$$

$$\dot{Q}_2 = 0.0909(287000-3500) + 0.3636(287000-56700)$$

$$\dot{Q}_2 = +109.5 \text{ kW}$$

$$\dot{Q}_3 = A_3\mathscr{F}_{3-1}(B_3-B_1) + A_3\mathscr{F}_{3-2}(B_3-B_2)$$

$$\dot{Q}_3 = 0.3636(57600-3500) + 0.3636(56700-287000)$$

$$\dot{Q}_3 = -64.4 \text{ kW}$$

The radiation network offers the designer a way of conceptualizing the radiation transfer process that is not afforded by the radiosity-irradiation formulation followed by matrix inversion. When the designer views the network representation he is given a grasp of how important various physical quantities such as surface emissivity are. For this reason the designer is urged to draw a network, perhaps with oversimplification to keep n small, to help him visualize the transfer problem, even when he may employ the matrix inversion algorithm to obtain numerical values.

3.E Selected Working Relations. With the radiation network as a working tool, it is an easy matter to obtain several useful results. Those reviewed are as follows:

1. A Source and a Sink.
    a) General Case.
    b) One Surface Not Concave.
    c) Parallel Walls.
    d) Concentric Tubes.
    e) Effective Emissivity of a Rough Surface.
2. A Source, a Sink, and a Refractory.
    a) General Case.
    b) Source and Sink Not Concave.
    c) Tube Bank over a Refractory.
3. Radiation Shields.
    a) General Case.
    b) Flat Shields.
    c) Concentric Cylindrical Shields.
    d) Concentric Spherical Shields.

The results here are subject to the assumptions of perfectly-diffuse surfaces and uniform radiosity. This latter assumption is exact for some geometries (parallel walls, concentric tubes) but only a rough approximation for others, such as short passages. Long passages, where the approximation of uniform radiosity is clearly inapplicable, are reviewed briefly in the next section.

3.E(a) A Source and a Sink. The general case of a single source and single sink forming a complete enclosure is represented by a simple series radiation network having merely a potential B_1, a surface resistance $R_1 = (1-\varepsilon_1)/\varepsilon_1 A_1$, a shape factor resistance $R_{1-2} = 1/A_1 F_{1-2}$, a second surface re-resistance $R_2 = (1-\varepsilon_2)/\varepsilon_2 A_2$, and potential B_2. Accordingly

$$\dot{Q}_1 = -\dot{Q}_2 = \frac{B_1 - B_2}{\frac{1-\varepsilon_1}{\varepsilon_1 A_1} + \frac{1}{A_1 F_{1-2}} + \frac{1-\varepsilon_2}{\varepsilon_2 A_2}} \quad (3\text{-}51)$$

$$A_1 \mathscr{F}_{1-2} = \frac{1}{\frac{1-\varepsilon_1}{\varepsilon_1 A_1} + \frac{1}{A_1 F_{1-2}} + \frac{1-\varepsilon_2}{\varepsilon_2 A_2}} \quad (3\text{-}52)$$

If one surface is not concave (denote it as number 1), then it cannot "see" itself; so $F_{1-1} = 0$ and $F_{1-2} = 1$. Equation (3-52) can be rearranged in the form

$$A_1 \mathscr{F}_{1-2} = \frac{A_1}{\frac{1}{\varepsilon_1} + \frac{A_1}{A_2}\left(\frac{1}{\varepsilon_2} - 1\right)} \quad (3\text{-}53)$$

For infinite parallel plates in addition $A_1 = A_2$, giving

$$A_1 \mathscr{F}_{1-2} = \frac{A_1}{\frac{1}{\varepsilon_1} + \frac{1}{\varepsilon_2} - 1} \quad (3\text{-}54)$$

For infinite concentric tubes, a length L has area $A_1 = \pi D_1 L$ where D_1 is understood to be the outside diameter of the inner tube, and $A_2 = \pi D_2 L$ where D_2 is the inside diameter of the outer tube. Equation (3-53) gives

$$A_1 \mathscr{F}_{1-2} = \frac{\pi D_1 L}{\frac{1}{\varepsilon_1} + \frac{D_1}{D_2}\left(\frac{1}{\varepsilon_2} - 1\right)} \qquad (3\text{-}55)$$

Equation (3.53) can be rearranged in the form of

$$\mathscr{F}_{1-2} = \frac{\varepsilon_1}{1 + \frac{\varepsilon_1 A_1}{\varepsilon_2 A_2}(1-\varepsilon_2)} \qquad (3\text{-}56)$$

Whenever $\varepsilon_1 A_1(1-\varepsilon_2)$ is small compared to $\varepsilon_2 A_2$, the transfer factor reduces to simply the emissivity of the small or non-black surface. This limit shows simply that a large space 2 can be regarded as a hohlraum, that is, as a black body. The presence of a sufficiently small object within it does not alter the black-body irradiation which falls upon a surface that sees only the hohlraum wall.

Consider a rough surface 1 within a large space 2. The surface roughness is gross compared to the wavelengths of concern. The large space is effectively a black body. If the surrounds are black and at isothermal temperature T_2, the area A_2 is of no concern; as shown by Eq. (3-56), the area A_2 disappears when $\varepsilon_2 = 1$. It is convenient to consider A_2 wrapped closely over the envelope of the surface roughness heights. The area A_2 is then the projected area of the surface. If the roughness heights are small compared to the radius of curvature of the mean surface, area A_2 can be regarded as flat, and surface A_1, because of its roughness, is concave. Accordingly it is F_{2-1} that is to be set to unity in Eq. (3-52), after substituting A_2F_{2-1} for A_1F_{1-2}, and ε_2 is set equal to 1. The result is

$$A_1\mathscr{F}_{1-2} = A_2\mathscr{F}_{2-1} = \frac{1}{\frac{1-\varepsilon_1}{\varepsilon_1 A_1} + \frac{1}{A_2}}$$

$$A_1\mathscr{F}_{1-2} = A_2\mathscr{F}_{2-1} = \frac{A_2}{\frac{A_2}{A_1}(\frac{1}{\varepsilon_1} - 1) + 1}$$

The quantity $\mathscr{F}_{2-1}$ is the effective emittance of the rough surface. It is defined so that the transfer factor $A_1\mathscr{F}_{1-2}$ is $A_2\varepsilon_{eff}$ when the surface is within a hohlraum. We have

$$\varepsilon_{eff} = \frac{1}{\frac{A_2}{A_1}(\frac{1}{\varepsilon_1} - 1) + 1} \qquad (3\text{-}57)$$

As the surface becomes smooth, and A_1 approaches A_2, the effective emittance becomes ε_1. As A_1 becomes large compared to A_2, the effective emittance goes to unity. The analysis is exact only for surface roughnesses formed by indentations that are segments of a sphere, because of the assumption of uniform radiosity, and is otherwise approximate. However, the qualitative result is true; the effective emissivity of a rough surface is bounded by ε_1 and unity and tends to increase with increasing A_1/A_2.

Occasionally a thermal designer who is unfamilar with the fundamentals of radiation heat transfer will decide to improve a radiator or heat exchanger by increasing A_1 in some novel manner to increase greatly $A_1\mathscr{F}_{1-2}$, even though subject to a fixed projected area A_2. But $A_1\mathscr{F}_{1-2} = A_2\mathscr{F}_{2-1}$ and $\mathscr{F}_{2-1}$ is bounded by emissivity ε_2, which equals unity at most. If A_2 is the projected area of rough surface A_1, there is no way to make $A_1\mathscr{F}_{1-2}$ exceed A_2, that is, to make ε_{eff} exceed unity.

3.E(b) A Source, a Sink, and a Refractory. The radiation network for a three-surface enclosure is shown in Figure 3-6. If surface 3 is a refractory, no heat crosses the m-surface, and this fact is represented by removing the current source shown as battery B_3. Node 3 is free floating, and resistance R_3 is of no concern, because no heat flow (current) flows through it. The equivalent conductance $A_1\mathscr{F}_{1-2}$ is given simply by

$$A_1\mathscr{F}_{1-2} = \cfrac{1}{R_1+R_2+\cfrac{1}{\cfrac{1}{R_{1-2}}+\cfrac{1}{R_{1-3}+R_{2-3}}}}$$

$$A_1\mathscr{F}_{1-2} = \cfrac{1}{\cfrac{1-\varepsilon_1}{\varepsilon_1 A_1}+\cfrac{1-\varepsilon_2}{\varepsilon_2 A_2}+\cfrac{1}{A_1F_{1-2}+\cfrac{1}{\cfrac{1}{A_1F_{1-3}}+\cfrac{1}{A_2F_{2-3}}}}} \tag{3-58}$$

This equation is clearly of the form

$$A_1\mathscr{F}_{1-2} = \cfrac{1}{\cfrac{1-\varepsilon_1}{\varepsilon_1 A_1}+\cfrac{1-\varepsilon_2}{\varepsilon_2 A_2}+\cfrac{1}{A_1\mathscr{F}_{1-2,b}}} \tag{3-59}$$

where $\mathscr{F}_{1-2,b}$ is the transfer factor accounting for the refractory when surfaces 1 and 2 are black,

$$A_1\mathscr{F}_{1-2,b} = A_1F_{1-2} + \cfrac{1}{\cfrac{1}{A_1F_{1-3}} + \cfrac{1}{A_2F_{2-3}}} \tag{3-60}$$

Eq. (3-59) is true whenever areas A_1 and A_2 can be assumed reasonably to have uniform radiosities, whether the refractory does or not. This point bears upon the long passages considered in the next section.

One often needs to know additionally B_3, in order to specify a refractory material suitable for the temperature T_3. One can proceed by finding the potentials q_1^+ and q_2^+, and the branch current $\dot{Q}_{1-3-2}$. With reference to Fig. 3-6 (B_3 open)

$$\dot{Q}_1 = A_1\mathscr{F}_{1-2}(B_1-B_2) \tag{3-61a}$$

$$q_1^+ = B_1-\dot{Q}_1R_1 \tag{3-61b}$$

$$q_2^+ = B_2+\dot{Q}_1R_2 \tag{3-61c}$$

$$\dot{Q}_{1-2} = \frac{1}{R_{1-2}}(q_1^+-q_2^+) \tag{3-61d}$$

$$\dot{Q}_{1-3-2} = \dot{Q}_1-\dot{Q}_{1-2} \tag{3-61e}$$

$$B_3 = q_1^+-\dot{Q}_{1-3-2}R_{1-3} \tag{3-61f}$$

When surfaces A_1 and A_2 are not concave, the simplification results that, since $F_{1-1} = 0$ and $F_{2-2} = 0$,

$$F_{1-3} = 1-F_{1-2}, \quad F_{2-3} = 1-F_{2-1} = 1-\frac{A_1}{A_2}F_{1-2}$$

Thus only the one shape factor F_{1-2} need be evaluated.

For example consider a cubical enclosure 1 m on a side having one facet being a source at $T_1 = 1500°K$, the floor a sink at $T_2 = 500°K$, and the remaining 3 walls and ceiling refractory at unknown temperature T_3. Suppose the surfaces are gray, and $\varepsilon_1 = \varepsilon_2 = 0.5$. What is the heat flow and the temperature of the refractory? To two-decimal-place accuracy, $F_{1-2} = 0.2$, so $F_{1-3} = F_{2-3} = 0.8$. From Eq. (3-60)

$$A_1\mathscr{F}_{1-2,b} = A_1F_{1-2} + \frac{1}{\dfrac{1}{A_1F_{1-3}} + \dfrac{1}{A_2F_{2-3}}}$$

$$\mathscr{F}_{1-2,b} = 0.2 + \frac{1}{\dfrac{1}{0.8} + \dfrac{1}{0.8}} = 0.60$$

Of the 80 per cent of the radiation leaving 1 and striking 3 half is reradiated back to 1 and the remainder to 2. This half, 40 per cent, exceeds the direct radiation, 20 per cent, by a factor of two.

To continue, the result for $\mathscr{F}_{1-2,b}$ is entered into Eq. (3-59) (with $A_1 = A_2 = 1\ m^2$) to obtain

$$\mathscr{F}_{1-2} = \frac{1}{\dfrac{1-0.5}{0.5} + \dfrac{1-0.5}{0.5} + \dfrac{1}{0.6}} = 0.273$$

Heat flow $\dot{Q}_1$ is found from Eq. (3-27)

$$\dot{Q}_1 = -\dot{Q}_2 = A_1\mathscr{F}_{1-2}(\sigma T_1{}^4 - \sigma T_2{}^4)$$

$$\dot{Q}_1 = (0.273)(5.6697\times10^{-8})(1500^4 - 500^4)$$

$$\dot{Q}_1 = (0.273)(287000-3500) = 77.3\ kW$$

To find B_3 we need not use Eq. (3-61), because of the symmetry in the network. From symmetry in this example we see at once that

$$B_3 = \frac{1}{2}(B_1+B_2) = 145300\ W/m^2$$

$$T_3 = (B_3/\sigma)^{1/4} = 1265°K$$

A tube or cylinder bank backed by a refractory wall to act as a reflector-reradiator is a configuration often used by the thermal design engineer. In one common application the cylinders are electrical heaters in rod form. In another, the tubes are filled with flowing fluid to be heated. The tubes are exposed to a furnace enclosure directly on one side. Consider the enclosure to be a hohlraum, and replace it with an imaginary black plane numbered 1. Number the tubes as 2 and the refractory back-up wall as 3 as shown in Fig. 3-8. The distance center-to-center is the pitch P, and the tube outside diameter is D. The tubes are separated to obtain more uniform heating or cooling of them around their periphery, that is, to make better use of the available heat transfer area to gain a higher heat flow per unit length of tube with specified temperature limits. If P were made equal to D, the back half of the tubes would not be utilized at all. It is desired to know the transfer factor $\mathscr{F}_{2-1}$ as a function of $r = P/D$.

If the tubes have a high emissivity (as they usually do) and if they are reasonably isothermal around their circumference, or if r is not too close to unity, their radiosity will be reasonably uniform, and they can be adequately represented by the single node 2. Similarly, if the refractory is spaced reasonably away from the tubes or if it is black and conducts laterally, it too can be represented by a single

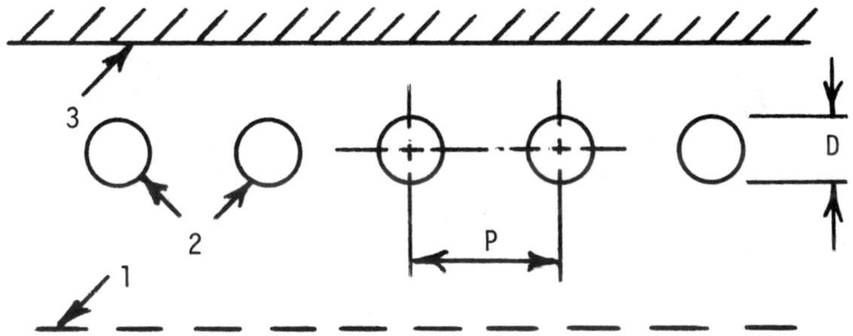

FIGURE 3-8

A Row of Circular Cylinders
Backed by a Refractory Wall

node. Then Eq. (3-58) applies directly. All that is necessary is to find the shape factors.

The string rule, Eq. (3-11) and Fig. (3-4), gives

$$F_{2-2} = \frac{2}{\pi}\left[(r^2-1)^{1/2}+\sin^{-1}(1/r)-r\right] \tag{3-62a}$$

Shape factor algebra determines the rest

$$F_{2-1} = F_{2-3} = (1-F_{2-2})/2 \tag{3-62b}$$

$$F_{1-2} = (A_2/A_1)F_{2-1} = (\pi/r)F_{2-1} \tag{3-62c}$$

$$F_{1-3} = 1-F_{1-2} \tag{3-62d}$$

For example, if P = 3D (r = 3), then we have the following values:

$$F_{2-2} = 0.1071 \;, \quad F_{2-1} = F_{2-3} = 0.4464$$

$$F_{1-2} = 0.4675 \;, \quad F_{1-3} = 0.5325$$

$$\mathcal{F}_{1-2,b} = 0.72$$

$$\mathcal{F}_{2-1,b} = 0.68$$

The value $\mathcal{F}_{2-1,b} = 0.68$ is 115 per cent greater than the value $1/\pi$ when the tubes are touching (r = 1), because the backs and sides of the tubes are better utilized. In the limit as r goes to infinity, $\mathcal{F}_{2-1,b}$ goes to unity, giving a 214 per cent increase over the $1/\pi$ value.

The normalized radiosity of the refractory can be obtained via Eq. (3-61), using values $B_1 = 1$ and $B_2 = 0$. The resulting value of radiosity is the irradiation on the point on the tube nearest the refractory wall, and this value can be compared with the value of unity pertaining to the point farthest from the wall to show the degree of uniformity in tube wall radiant heat flux.

The quantity $\mathscr{F}_{2-1,b}$ is easily calculated from Eqs. (3-60 and 62) by hand. Figure 3-9 shows the behavior vs. r. To two decimal places

$$\mathscr{F}_{2-1,b} \doteq 1 - (1/r) \qquad r \geq 4$$

Hottel and Sarofim [8] show additional results for double rows of tubes in front of a refractory wall.

3.E(c) Radiation Shields. Radiation shields are used by the thermal designer in various forms: aluminized plastic sheet, aluminum-foil-paper laminate, thin stainless steel sheet, ceramic tubes, to name a few. The purpose is to reduce unwanted heat transfer. At high temperatures or under vacuum conditions, conduction may be negligible (Section 8 considers combined mode transfer). The radiation shield then can be represented on a radiation network as a node with a free-floating radiosity B with a surface resistance on either side. The radiation shield is a two-sided refractory. Consider an array of N shields of the same material between an inner black source area A_1 and a black sink area A_{N+2}. As before, surface resistances to account for nonblack source and sink can be easily added as shown in Eq. (3.59). Between N shields there are N-1 spaces, each spacing having resistance

FIGURE 3-9

Tube-to-Furnace Transfer Factor for
Single Tube Bank Over Refractory Wall

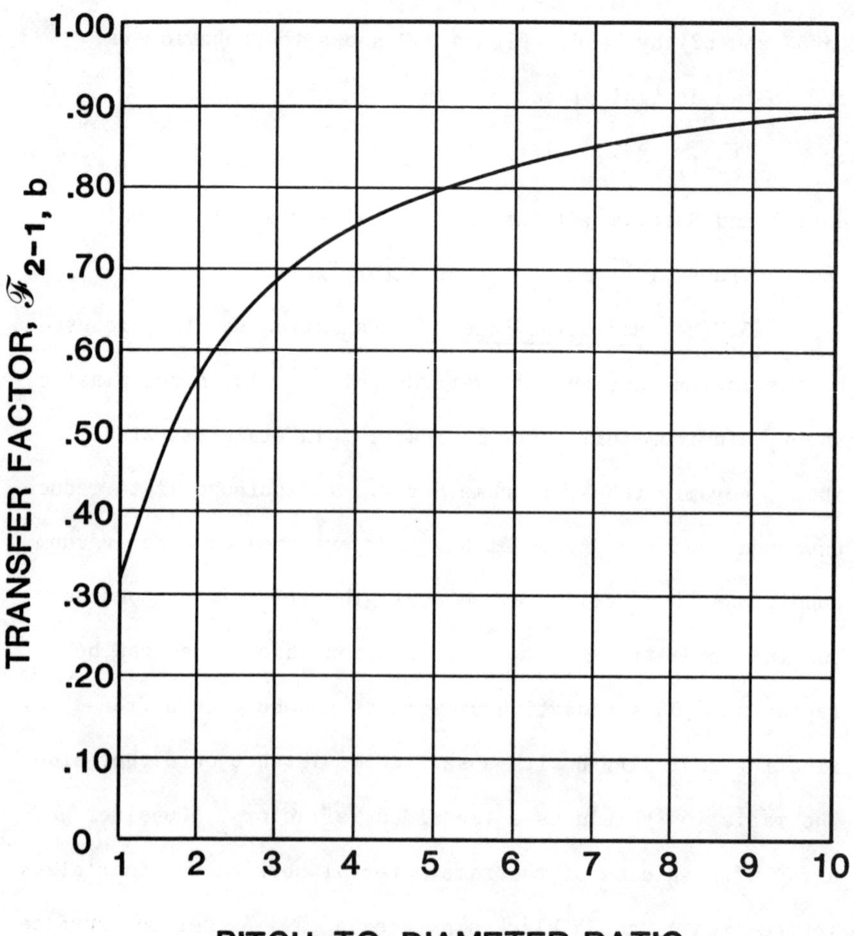

$$R_{s,n} = \frac{1-\varepsilon_f}{\varepsilon_f A_{f,n}} + \frac{1}{A_{f,n}} + \frac{1-\varepsilon_b}{\varepsilon_b A_{b,n+1}}, \quad n=2 \text{ to } N \qquad (3\text{-}63)$$

where subscript f refers to the front (or outside) of the smaller shield and b to the back (or inside) of the larger one. In addition to N-1 of these resistances there is the surface resistance of the back of the inner shield, the shape factor resistance from the inner surface 1, the surface resistance of the front of the outer shield N+1 and its shape factor resistance. Therefore

$$A_1 \mathscr{F}_{1-(N+2),b} = \frac{1}{\frac{1}{A_1} + \frac{1-\varepsilon_b}{\varepsilon_b A_{b,2}} + \sum_{n=2}^{N} R_{s,n} + \frac{1-\varepsilon_f}{\varepsilon_f A_{f,N+1}} + \frac{1}{A_{f,N+1}}} \qquad (3\text{-}64)$$

For flat shields $A_1 = A_{f,n} = A_{b,n}$. The areas cancel, and

$$\mathscr{F}_{1-(N+2),b} = \frac{1}{N\left[\frac{1}{\varepsilon_f} + \frac{1}{\varepsilon_b} - 1\right] + 1} \qquad (3\text{-}65)$$

In this case the transfer factor falls with 1/N as N gets large.

For concentric cylindrical shields with constant front-to-front spacing s, thickness δ, and inner source or sink radius r_1, there holds

$$A_1 = 2\pi r_1 L \qquad (3\text{-}66a)$$

$$A_{f,n} = 2\pi(r_1+ns)L \qquad (3\text{-}66b)$$

$$A_{b,n} = 2\pi(r_1+ns-\delta)L \qquad (3\text{-}66c)$$

When s/r_1 is small, the summation can be approximated with an integral, and the transfer factor falls as $1/\ell n N$ as N gets large.

For concentric spherical shields the following relations pertain:

$$A_1 = 4\pi r_1^2 \tag{3-67a}$$

$$A_{\delta,n} = 4\pi(r_1+ns)^2 \tag{3-67b}$$

$$A_{b,n} = 4\pi(r_1+ns-\delta)^2 \tag{3.67c}$$

When s/r_1 is small, the summation in Eq. (3-64) can be approximated as an integral, and as N gets large the additional shields produce no added benefit in reducing the transfer factor.

For small s/r_1 the concept of radiation conductivity can be used, and that conductivity can be combined with simultaneous gaseous conduction as explained in Section 8 to obtain an approximate simple closed-form result more convenient than Eqs. (3-64 and 66 or 67).

3.F <u>Diffuse-Walled Passages</u>. The thermal designer may wish to assess the importance or unimportance of axial radiation, for example, in a regenerative heat exchanger used in a Brayton- or Stirling-cycle engine or in a combustion air preheater. The leakage of heat radiation through a hole or crack penetrating through thermal insulation is a common concern. In what follows shape factor algebra is used to formulate the heat flux down a passage. When the side wall radiosity is known, as it is for a black-walled passage with known temperature distribution, or when the side-wall radiosity can be reasonably approximated, as for the refractory-walled passage, the formulation is of direct practical use. When the radiosity is unknown and cannot be readily approximated, the problem is easily formulated using the method illustrated here, and numerical methods of solution can be employed. However, modern practice is to use a direct Monte Carlo calculation. The Monte Carlo technique is described in Section 4.

Consider a passage of length L in the form of a right cylinder with arbitrary cross-section. Denote the ends as 1 and 2 and the side walls as 3. Let x run from end 2 towards 1. Consider an imaginary surface 4 located at $x = x_o$. The dimensionless irradiation upon the side of the surface facing end 1 is given by Eq. (3-14), generalized by passing from the sum to an integral.

$$q_4^{*-} = F_{4-1}q_1^{*+} + \int_{x=x_o}^{x=L} q_3^{*+}(x) \frac{dF_{4-3}}{d(x-x_o)} d(x-x_o)$$

The shape factor F_{4-1} is of the same form as $F_{1-2}(L)$, but the length in question is $L-x_o$; that notation is adopted. Shape factor algebra shows that

$$F_{4-3}(x-x_o) = 1-F_{4-1}(x-x_o) = 1-F_{1-2}(x-x_o)$$

Accordingly the equation may be written

$$q_4^{*-}(x_o) = F_{1-2}(L-x_o)$$
$$+ \int_{x=x_o}^{x=L} q_3^{*+}(x) \frac{d}{d(x-x_o)} [1-F_{1-2}(x-x_o)]\, d(x-x_o) \qquad (3\text{-}68)$$

The dimensionless radiosity, irradiation, and flux are understood to be

$$q_n^{*\pm} = \frac{q_n^{\pm}-q_2^{+}}{q_1^{+}-q_2^{+}}, \quad q^* = \frac{q}{q_1^{+}-q_2^{+}} \qquad (3\text{-}69)$$

Note that q_1^{*+} is unity and q_2^{*+} is zero, so that they do not appear explicitly in Eq. (3-68) or Eq. (3-70) to follow.

Let surface 5 be the back of surface 4, that is, the side of the cross-section at $x=x_o$ facing end 2. Then the irradiation upon 5 is given by an expression similar to that for 4.

$$q_5^{*-} = \int_{x=x_o}^{x=0} q_3^{*+} \frac{d}{d(x_o-x)} [1-F_{1-2}(x_o-x)]\, d(x_o-x)$$

But the irradiation upon 5 is the radiosity of 4. The net heat flux *in the negative-x direction* is from Eq. (3-12),

$$q^*(x_o) = F_{1-2}(L-x_o) + \int_{x=x_o}^{x=L} q_3^{*+} \frac{d}{d(x-x_o)} [1-F_{1-2}(x-x_o)] d(x-x_o)$$

$$- \int_{x=x_o}^{x=0} q_3^{*+} \frac{d}{d(x_o-x)} [1-F_{1-2}(x_o-x)] d(x_o-x) \tag{3-70}$$

Equation (3-70) is the general expression for the average heat flux at any location x_o for a diffuse-walled passage having ends with uniform radiosity. Its utility is made more clear by considering the special case when

$$q_3^{*+}(x) = mx+b \tag{3-71}$$

The reader recognizes this equation as the intercept-slope representation of a straight-line side-wall radiosity distribution.

Consider the first integral in Eq. (3-70); denote it as I_1. Let $u = x-x_o$

$$I_1 = \int_0^{L-x_o} [mx+b+mx_o-mx_o] \frac{d}{du} [1-F_{1-2}(u)]du$$

$$I_1 = [b+mx_o][1-F_{1-2}(L-x_o)] + m\int_0^{L-x_o} u \frac{d}{du} [1-F_{1-2}(u)]du$$

Integrate the remaining integral by parts to obtain

$$I_1 = [b+mL][1-F_{1-2}(L-x_o)] - m[L-x_o] + m[L-x_o]S(L-x_o)$$

where

$$S(L) \equiv \frac{1}{L}\int_0^L F_{1-2}(x)dx \tag{3-72}$$

When the same procedure is followed with the second integral, Eqs. (3-70 and 71) yield

$$q^*(x_o) = [1-(b+mL)]F_{1-2}(L-x_o)+bF_{1-2}(x_o)$$

$$+ m(L-x_o)S(L-x_o)+mx_oS(x_o) \qquad (3\text{-}73)$$

When one recalls that q_1^{*+} is unity, q_2^{*+} is zero, $q_3^{*+}(0) = b$, and $q_3^{*+}(L) = b+mL$, then the symmetry in the equation with respect to x_o and $L-x_o$ may be appreciated.

The cross-sectional geometry fixes the relations for F_{1-2} and S. For the slot with height H = 1 (i.e. L is L/H = $2L/D_H$, where D_H is the hydraulic diameter, and x = x/H = $2x/D_H$); the string rule shows

$$F_{1-2}(x) = (1+x^2)^{1/2}-x \qquad (3\text{-}74a)$$

$$S(x) = (1/2)((1+x^2)^{1/2}-x)+(1/2x)\ln(x+(1+x^2)^{1/2})$$

$$(3\text{-}74b)$$

For the circle with diameter D (i.e. L=L/D and x=x/D),

$$F_{1-2}(x) = 1+2x^2-2x(1+x^2)^{\frac{1}{2}} \qquad (3\text{-}75a)$$

$$S(x) = 1+(2/3x)(1+x^3-(1+x^2)^{3/2}) \qquad (3\text{-}75b)$$

For example, suppose one wanted to estimate the radiant heat flux into the passages of a 5 cm long regenerator core having approximately circular (D_h = 5 mm) passages of a high emissivity material. Suppose the inlet to the hot side of the regenerator is approximately a hohlraum at T_1 = 600°K, and the outlet is one at T_2 = 350°K. The wall itself is T_L =

575°K at the inlet and T_o = 325°K at the outlet. It is agreed approximate the radiosity distribution along the passage wall as a straight line from σT_o^4 to σT_L^4. The net heat flux into a passage at its inlet (at x_o= L = 5 cm) is desired.

As a first step Eq. (3-73) is written for x_o= L

$$q^*(L) = 1-(b+mL)+bF_{1-2}(L)+mLS(L)$$

The values of b and mL are found from Eq. (3-69)

$$b = q_3^{*+}(0) = \frac{\sigma T_o^4 - \sigma T_2^4}{\sigma T_1^4 - \sigma T_2^4} = \frac{325^4 - 350^4}{600^6 - 350^4} = -0.034$$

$$mL = q_3^{*+}(L) - q_3^{*+}(0) = \frac{\sigma T_L^4 - \sigma T_o^4}{\sigma T_1^4 - \sigma T_2^4} = \frac{575^4 - 325^4}{600^4 - 350^4} = 0.857$$

$$b+mL = q_3^{*+}(L) = \frac{\sigma T_L^4 - \sigma T_2^4}{\sigma T_1^4 - \sigma T_2^4} = \frac{575^4 - 350^4}{600^4 - 350^4} = 0.823$$

Then the values of F_{1-2} and S are found for L = 5/0.5 = 10 from Eqs. (3-75a and b), and q*(L) is found.

$$F_{1-2}(10) = 0.0025$$

$$S(10) = 0.0642$$

$$q^* = 1-0.823-0.034(0.0025)+0.857(0.0642)$$

$$q^* = 0.232$$

The dimensional heat flux follows from Eq. (3-69)

$$q = (\sigma T_1^4 - \sigma T_2^4)q^* = (5.6697\text{x}10^{-8})(600^4-350^4)(0.232)$$

$$q = 1507 \text{ W/m}^2$$

When the side wall radiosity is unknown but the temperature or black-body radiosity variation with x is known, the problem can be formulated by passing from the sum to the integral in Eq. (3-14) and using Eq. (3-13) to eliminate q^+ or q^-. The resulting linear integral equation can be solved numerically. Alternatively, the side wall can be divided into a number of segments, and Eq. (3-23 or 25) can be used and solved by matrix inversion as described previously. Shape factor algebra allows all the shape factors to be expressed and calculated in terms of differences of $F_{1-2}(x_i - x_j)$. However, as described in the next section, a Monte Carlo algorithm is more easily applied.

In the case of refractory side walls, the temperature or black-body radiosity variation with x is unknown. All that is known is that $q_3^+(x) = q_3^-(x)$ regardless of side wall emissivity. An equivalent statement is simply that

$$\frac{d}{dx_o} q^*(x_o) = 0 \qquad (3\text{-}76)$$

where $q^*(x_o)$ is given by Eq. (3-70). Note that when conduction is absent no boundary conditions can be imposed. Again the side wall can be divided into a number of zones, the integrals approximated as sums, and matrix inversion applied. Again Monte Carlo calculations are easily programmed for computer execution. What is desired in the end is q^*, which is the $\mathscr{F}_{1-2,b}$ needed for Eq. (3-59) to find the transfer factor $\mathscr{F}_{1-2}$.

An approximate closed-form expression for the refractory-walled-passage transfer factor $\mathscr{F}_{1-2,b}$ can be found by postulating q_3^{*+} to be a straight line passing through 1/2 at $x = L/2$. The unknown slope m is found, following Hottel and Kellar [19], by requiring that Eq. (3-73) give the same result for $x_0 = 0$ and $x_o = L/2$ (also $x_o = L$ from symmetry). There is found

$$mL = \frac{G(L)-F(L/2)}{[G(L)-F(L/2)]+[S(L/2)-S(L)]} \tag{3-77}$$

where

$$G(L) = \frac{1+F(L)}{2}, \quad F(L) = F_{1-2}(L) \tag{3-78}$$

The value of the transfer factor is

$$\mathscr{F}_{1-2,b} = (1-mL)\, G(L)+mLS(L) \tag{3-79}$$

For the circle cross-section, the notion of an extrapolation length E is a good approximation, i.e.,

$$mL \doteq \frac{L}{L+2E}, \quad E = 0.5 \text{ for circle}$$

Note that if conduction from the hot source to the end of the passage at $x = L$ and from the end at $x = 0$ to the cold sink is operative, E should be zero and $mL = 1$.

Figure 3-10 shows values presented versus the length-to-hydraulic-diameter ratio, because transfer factor is independent of cross-sectional shape for small L/D_H. For small L, Eq. (3-60 or 79) indicates

$$\mathscr{F}_{1-2,b} = \frac{1+F_{1-2}}{2} \quad \text{as} \quad L \to 0$$

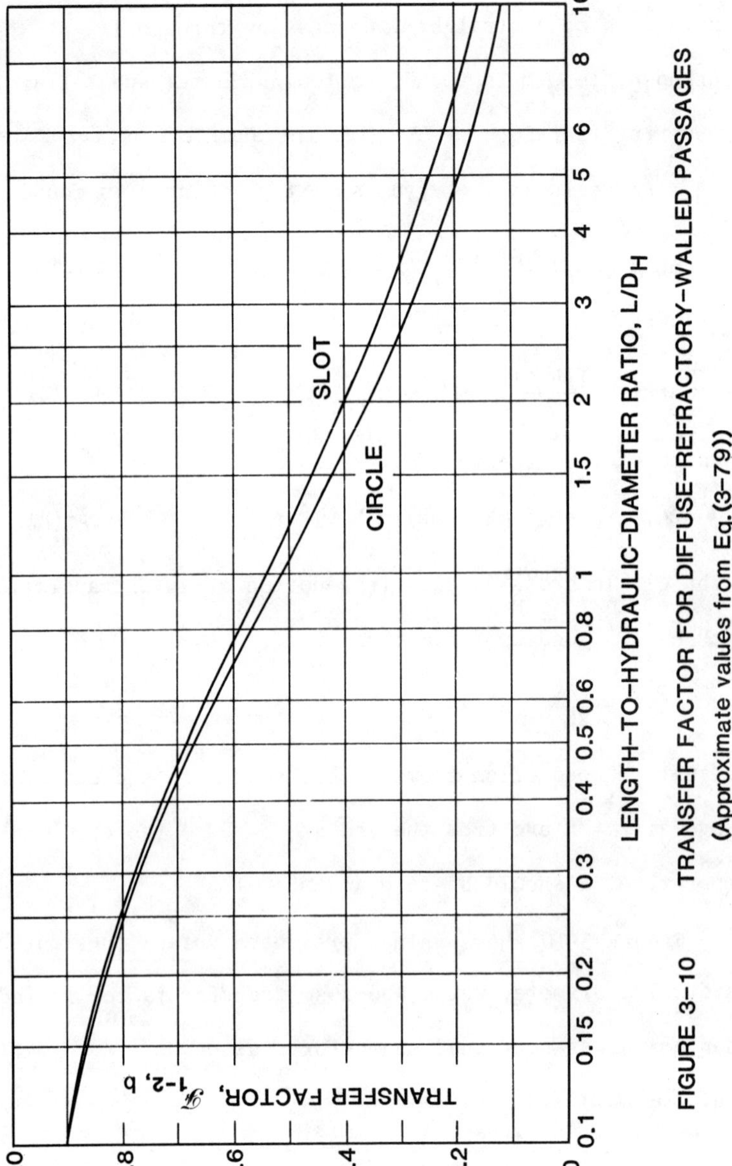

FIGURE 3–10 TRANSFER FACTOR FOR DIFFUSE–REFRACTORY–WALLED PASSAGES
(Approximate values from Eq.(3–79))

But shape factor algebra shows

$$F_{1-2} = 1-F_{1-3} = 1-(A_3/A_1)F_{3-1}$$

Area A_3 is PL where P is the perimeter of the passage cross-section. When L is small F_{3-3} is small so that

$$F_{3-1} = F_{3-2} = \frac{1-F_{3-3}}{2} \doteq \frac{1}{2}$$

Accordingly, for small L, but regardless of cross-section, one has

$$\mathscr{F}_{1-2,b} \doteq 1 - \frac{PL}{4A_1} = 1 - \frac{L}{D_H}$$

$$\mathscr{F}_{1-2,b} \doteq \frac{1}{1+(L/D_H)}, \quad L/D_H \text{ small} \qquad (3\text{-}80)$$

Figure 3-10 shows that the insensitivity of the transfer factor persists even to values of L/D_H considerably larger than unity.

For example, suppose one wanted to know the heat leak per meter ($\dot{Q}/W$) through a 3 mm crack in a 6 cm thick insulation layer. Suppose T_1 = 600°K and T_2 = 300°K. One finds the transfer factor from the figure.

$$D_H = 4A/P = 4WH/2W = 2H = 0.6 \text{ cm}$$

$$L/D_H = 6/0.6 = 10, \quad \mathscr{F}_{1-2,b} \doteq 0.16$$

Then one uses the definition of the transfer factor, Eq. (3-27), to evaluate the heat flow.

$$\dot{Q}/W = qWH/W = Hq = H\mathscr{F}_{1-2,b}\sigma(T_1^4-T_2^4)$$

$$\dot{Q}/W = (0.006)(0.16)(5.6697\text{x}10^{-8})(600^4-300^4)$$

$$\dot{Q}/W = 6.6 \text{ W/m}$$

Values of $\mathscr{F}_{1-2,b}$ are available in the literature for a few other cross-sections and for tapered passages [21,22]. Specular-walled passages and passages modeled by Torrance-Sparrow wall roughness model are discussed in Section 4.

REFERENCES TO SECTION 3

I. Shape Factors

A. Definition, Use, and Catalogs

1. Moon, P., The Scientific Basis of Illuminating Engineering, McGraw Hill, New York, 1936, Chapter X.

2. Hamilton, D. C., and W. R. Morgan, "Radiant Interchange Configuration Factors", NACA TN 2836, 1952.

3. Hottel, H. C., "Radiant Heat Transmission", Chapter 4 of W. H. McAdams' Heat Transmission, McGraw Hill, New York, 3rd Edition, 1954.

4. Zijl, H., Large-Size Perfect Diffusers, Philips' Gloeilampenfabrieken, Eindhoven, Netherlands, 2nd Revised Edition, 1960.

5. Siegel, R., and J. R. Howell, Thermal Radiation Transfer, McGraw Hill, New York, 1972 (see Appendix C, "Catalog of Selected Configuration Factors").

B. Optical and Mechanical Devices

6. Jakob, M., Heat Transfer, Volume II, John Wiley, New York, 1957, pp. 66-85.

7. Hickman, R. S., "The Measurement of Radiation Configuration Factors with Parabolic Mirrors", Proceedings Heat Transfer and Fluid Mechanics Institute, 1962, Stanford University Press, pp. 89-94.

8. Farrell, R., "Determination of Configuration Factors of Irregular Shape", J. Heat Transfer, Vol. 98, pp. 311-313, 1976.

C. Computer Programs

9. Dummer, R. S., and W. T. Breckenridge, Jr., "Radiation Configuration Factors Program", Report ERR-AN-224, General Dynamics Astronautics, 1963.

10. Toups, K. A., "CONFAC II", Technical Documentary Report RDL-TDR-64-43, North American Aviation, 1964.

II. Radiosity-Irradiation Formulations

11. Hottel, H. C., and A. F. Sarofim, Radiative Transfer, McGraw Hill, New York, 1967, See also Ref. [3].

12. Gebhart, B., "Unified Treatment for Thermal Radiation Transfer Processes--Gray, Diffuse Radiators and Absorbers", ASME Paper 57-A-34 (1957).

13. O'Brien, P. F., "Lightning Calculations for Thirty-Five Thousand Rooms", Illuminating Engineering, Vol. 55, pp. 215-226, 1960.

14. Bobco, R. P., "Radiation Heat Transfer in Semigray Enclosures with Specularly and Diffusely Reflecting Surfaces", J. Heat Transfer, Vol. 86, pp. 123-130, 1964.

15. Ishimoto, T., and J. T. Bevans, "Method of Evaluating Script F for Radiant Exchange within an Enclosure", AIAA Journal, Vol. 1, pp. 1428-1429, 1963.

16. Wiebelt, J. A., Engineering Radiation Heat Transfer, Holt, New York, 1966.

III. Radiation Network

17. Openheim, A. K., "Radiation Analysis by the Network Method", Trans. Am. Soc. Mech. Engrs., Vol. 78, pp. 725-735, 1956.

18. Oppenheim, A. K., "The Engineering Radiation Problem--An Example of the Interaction Between Engineering and Mathematics", Zeitschrift fur angewandte Mathematik und Mechanik, Vol. 36, pp. 81-93, 1956.

IV. Diffuse Passages

19. Hottel, H. C., and J. D. Keller, "Effect of Reradiation on Heat Transmission in Furnaces and Through Openings", Trans. Am. Soc. Mech. Engrs., Vol. 55, pp. 39-49, 1933.

20. Perlmutter, M., and R. Siegel, "Effect of Specularly Reflecting Gray Surface on Thermal Radiation through a Tube and from its Heated Wall", J. Heat Transfer, Vol. 85, pp. 55-62, 1963.

21. Sparrow, E. M., and V. K. Jonsson, "The Transport of Radiant Energy through Tapered Tubes and Gaps", J. Heat Transfer, Vol. 86, p. 132, 1964.

22. Kezios, S. P., and W. Wulff, "Radiative Heat Transfer Through Openings of Various Cross Sections", Proc. Third International Heat Transfer Conf., AIChE, New York, Vol. 5, 1966, pp. 207-218.

EXERCISES FOR SECTION 3

1. Find the shape factor from a desk top to a window, assuming the window is a diffuse source. With one lower corner of the room taken to be (0,0,0), the desk lies in the plane z = 73 and the x-y region, 90 < x < 240 and 105 < y < 180 cm. The window lies in the plane x = 0 in the y-z region 103 < z < 223 and 0 < y < 180 cm.

2. A long tunnel furnace has the load on a floor-mounted conveyer belt and is heated by a back-wall muffle. The ceiling and front wall are unheated. The cross-section is square; the furnace brick is believed to have an emissivity of 0.55, and the load one of 0.70. A total pyranometer is sighted on the walls with the following results:

   Ceiling: T = 1150°C

   Back wall: T = 1340°C

   Front wall: T = 1140°C

   Load: T = 830°C

   a) What are the true temperatures of the four surfaces?

   b) What is the net heat flux at each surface?

   c) How much does the performance of the furnace depart from a source-sink-refractory model?

3. An industrial furnace is 2.4 m by 2.4 m in cross-section and 6 m long. The two 2.4x6 m side walls are heated to 1400°C. The load on the floor is 800°C. The ceiling

3. (continued)

and ends are refractory. The side wall emissivities are 0.70, and the load emissivity is 0.50.

a) Find the shape factors F_{1-2}, F_{1-R}, and F_{2-R} where 1 is the source, 2 the load, and R the refractory.

b) Find the transfer factor $\mathscr{F}_{1-2}$ including the effect of the refractory.

c) What is the heat transfer $\dot{Q}$ in kW?

d) What apparent temperature would a total pyranometer indicate when aimed at the source, the ceiling, or the load?

e) If the load were covered with a high emissivity coating, $\varepsilon = 0.90$, how much would $\dot{Q}$ be increased?

4. A furnace similar to that above (2.4 x 2.4 x 6 m) is heated by an array of vertical resistance-heated rods along one side wall. The remaining side, ends, and ceiling are refractory. The rods are 2.5 cm in diameter, pitched 7.5 cm center-to-center, and have an emissivity of 0.7. Their maximum allowable temperature is 1150°C. The load on the floor is 800°C with an emissivity of 0.50. Find the maximum allowable $\dot{Q}$ in kW.

5. Prove that the diffuse-walled sphere has a constant value of irradiation over its entire interior, even when T varies arbitrarily over its surface.

6. A hollow spherical spacecraft with tangentially-poorly-conducting but radially-well-conducting walls is continuously exposed to solar radiation $I^{-}d\Omega = 1380\ W/m^2$ on its sunny-side exterior. The uniform exterior solar absorptivity varies with angle of incidence according to $\alpha_e = \alpha_o[0.5\sec\theta + 0.5\cos\theta]$ with $\alpha_o = 0.08$, and the total hemispherical emissivity of the exterior is $\varepsilon_i = 0.10$. The internal total emissivity is 0.8 over the entire interior wall.

   a) Write the heat balance equation giving $\sigma T^4(\theta)$, assuming the internal as well as the external radiation is known.

   b) Write the radiosity-irradiation equation for the interior (see exercise 5), and substitute the heat balance expression for $\sigma T^4(\theta')$ into it to obtain one equation in q_i^-. Find the irradiation, $T(\theta=0)$, $T(\theta=89.9°)$, and $T(\theta=180°)$.

7. An integrating sphere is 10 cm in radius. It contains on its wall a 1 cm^2 detector with an absorptivity of 0.90, and a 1x2 cm entrance port opening into a blackbody cavity. The remaining portion of the sphere is smoked with magnesium oxide having a short wavelength reflectance of 0.98. Radiant energy enters the sphere through the entrance port onto an element of the wall and then is reflected and interreflected, giving rise to

7. (continued)

a sphere wall radiosity and irradiation. Thermal radiation is not detected because the source radiation is chopped, and the detector-amplifier system responds only to the chopped radiation. Find the fractions of the chopped incoming radiation that are (a) lost out the entrance port, (b) absorbed by the MgO-smoked wall, and (c) absorbed by the detector. (Item (c) is called the "sphere efficiency".) Contrast the ease of solution from the radiosity-irradiation approach (in the light of Exercise 5) with that from the network approach.

8. A room is very roughly modeled as a spherical enclosure with 100 m^2 of surface area. An irradiation of 10 W/m^2 is to be provided at night by fluorescent lights.

   a) Give an equation showing how the reflectivity of the room paint, furnishings, and drapes affects the amount of light required.

   b) If the average reflectivity of all the surfaces in the room is 0.60, how many Watts are required when the drapes are closed?

   c) How many Watts are required when the drapes are open, exposing 10 m^2 of window having a reflectivity of only 0.07?

9. Develop a network representation for diffuse reflection of solar energy by deriving what the network voltage source should be in the case of direct irradiation. For example, suppose that open Venetian blinds (parallel white slats of length L and spacing D) reflect diffusely with a reflectivity of 0.65 and have solar irradiation incident upon them at 45° from their axis. If $Id\Omega\cos\theta =$ 600 W/m^2 is incident upon the tops of the slats with L/D = 1, what is the irradiation onto the openings into the room? Neglect thermal emission.

10. A light well is used to make the inhabitant of a house put underground for thermal storage think that he is above ground. A well 2 m in diameter is 6 m deep and has its sides and bottom plastered with white stucco (reflectivity 0.90). Centered about depths of 0.5, 3.0 , and 5.5 m are triple glazed windows 1 m high and 0.5 m wide having a solar reflectivity of 0.18. As an approximation it is agreed to distribute the effect of the windows by adjusting the side wall reflectivity to an area-weighted mean value. Solar radiation of $Id\Omega$ = 1000 W/m^2 enters the top of the well directly at 30° from the horizontal and indirectly by reflection from a 2.5 m high long stucco vertical wall. It is agreed to distribute the radiation entering the top of the well uniformly over the top 2 m increment of the well's side wall.

10. (continued)

Using a five-node approximate analysis (1 is the black well opening; 2, 3, and 4 are 2-m high side elements, and 5 is the bottom), set up the radiosity-irradiation equations and solve by iteration. Find the radiosities of the three side elements and the bottom of the well.

11. A space vehicle in the form of a cube is oriented with one side always facing the sun and the others in the shade. It is agreed that all interior and exterior surfaces must be one of the following: (a) Black Paint $\alpha_S = 0.95$, $\varepsilon = 0.86$; (b) White Paint, $\alpha_S = 0.13$, $\varepsilon = 0.86$; (c) First Surface Aluminized Tape, $\alpha_S = 0.10$, $\varepsilon = 0.03$. Recommend surface finishes so that the shaded sides operate at 10°C and the sunlit side is as cool as possible. Assume each face of the cube is isothermal, but neglect conduction from one face to another.

12. A seat-of-the-pants engineer approximates $A_1 \mathscr{F}_{1-2}$ with $\varepsilon_1 \varepsilon_2 A_1 F_{1-2}$ for calculating radiation heat transfer between any two surfaces 1 and 2 in an enclosure. Give a brief critique of this approximation. Consider the following points in your critique:

a) What is clearly a lower limit to $A_1 \mathscr{F}_{1-2}$?

b) What is clearly an upper limit to $A_1 \mathscr{F}_{1-2}$?

c) What is the exact value of $A_1 \mathscr{F}_{1-2}$ when all surfaces including 1 and 2 lie on the interior of a diffuse

12. (continued)

walled sphere?

d) How much might $A_1 \mathscr{F}_{1-2}$ differ from $\varepsilon_1 \varepsilon_2 A_1 F_{1-2}$ when areas 1 and 2 are the ends of a long circular passage with refractory side walls? How does $A_1 \mathscr{F}_{1-2}$ vary with L/D when L/D is large? How does $A_1 F_{1-2}$ vary then?

13. Calculate $\mathscr{F}_{1-2,b}$ vs. L/D_h for transmission through a diffuse-refractory-sided passage with 2 to 1 rectangular cross-section. Plot your results for $1 \leq L/D_h \leq 8$ on a copy of Fig. 3-10 of the Notes.

14. A plastic film is used instead of glass for the inner glazing in a double-glazed solar collector. Assume that the outer coverglass is opaque, diffuse, and gray with an emissivity of 0.80 and a temperature of 300 K. Assume that the solar absorber plate is opaque and diffuse at a temperature of 400 K, but is nongray with an emissivity of 0.90 for $\lambda < 1.0$ μm, 0.25, $1 < \lambda < 10$ μm, and 0.10, $\lambda > 10$ μm. Assume that the plastic film has the following radiation characteristics

$\rho = 0.04$, $\tau = 0.96$ $\lambda < 1$ μm

$\rho = 0.02$, $\tau = 0.48$ $1 < \lambda < 10$ μm

$\rho = 0.03$, $\tau = 0.74$ $10 < \lambda$ μm

a) Write the spectral radiosity-irradiation equations for the situation where the plastic film has a black

14. (continued)

body radiosity value of unity and the others have zero values.

b) Solve algebraically for the spectral transfer factors from the film to the other two elements.

c) Assuming the plastic film has a temperature of say 350 K, find the total transfer factors from the film to the other two elements.

4 RADIATION TRANSFER BETWEEN NONDIFFUSE WALLS

4.A Specular and Imperfectly Diffuse Surfaces. The concept of a perfectly diffuse surface was an artifice introduced to simplify formal mathematical analysis of radiant transfer. The concept is widely used in engineering design and analysis for its convenience, and, surprisingly, the answers so obtained are found in many instances to be remarkably close to the answers found with more realistic analytical models. There are situations, for example in transmission through long passages with specular side walls, where the assumption of perfectly diffuse reflection will lead to serious error, however. Thus the designer or analyst needs to be able to carry out calculations when one or more surfaces are not pefectly diffuse.

Perfectly diffuse reflection is where the bidirectional reflectance is a constant independent of all four angles, the two angles of incidence and the two of emergence. The antithesis of perfectly diffuse reflection is specular reflection where the bidirectional reflectance is identically zero for all directions of emergence except the specular angle where it has an integrable singularity. Imperfectly diffuse is the term usually used to denote that the bidirectional reflectance is nonzero but not constant with angles of emergence. Mixed specular diffuse reflection occurs when there is a specular component, for example, from the smooth surface of the binder of a glossy enamel paint, and a (perfectly or imprefectly) diffuse component, for example, from the underlying particles of pigment of such a glossy enamel.

4.B The Mirror Image Concept. The mirror-image concept was introduced formally into thermal radiation transfer analysis by Eckert, Sparrow, and coworkers [1,2]. The concept is useful primarily when the enclosure in question contains only a few plane specular surfaces arranged so that the number of multiple specular reflections is either limited or forms an easily summed chain. The concept is based upon the fact that a ray coming from an element of diffuse surface i and reflected by mirror m to an element of diffuse surface j can be regarded as an uninterrupted straight line from i to the mirror image of j. Thus the shape factor of Eq. (3.5) can be applied between surface i and the mirror image of j in calculating the transfer between i and j via m. The image of j as seen in m is denoted j(m). The mirror-image shape factor is then written $F_{i-j(m)}$.

If another mirror n is present, the mirror image of j as seen via m may also be viewed via n. Thus there is a mirror image in n of the mirror image in m of surface j. Figure 4-1 illustrates the single and multiple image concept. Use of the mirror image concept involves assuming that the specular reflectivity of the mirrors does not vary

FIGURE 4-1

The Mirror-Image Concept

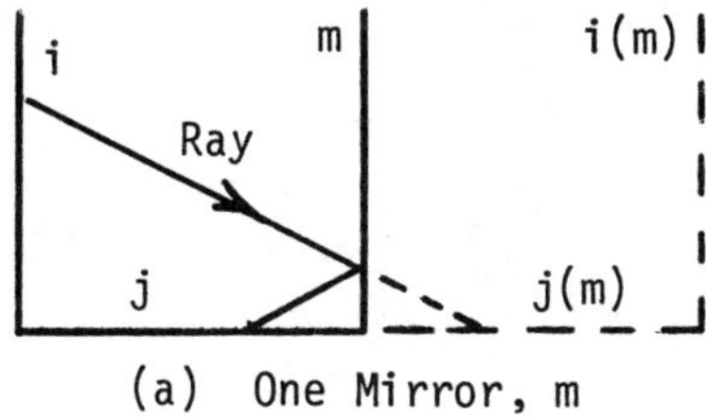

(a) One Mirror, m

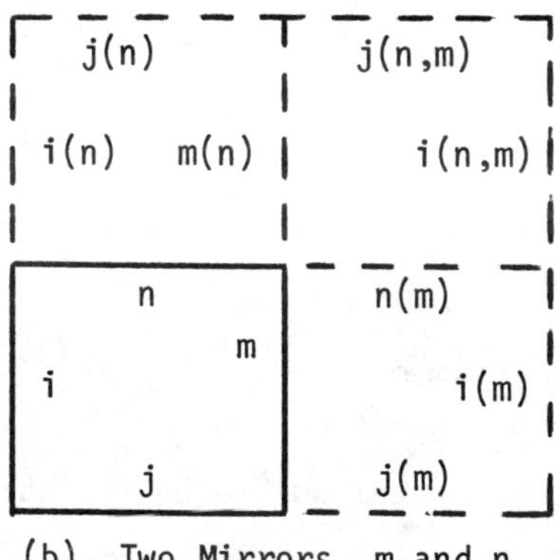

(b) Two Mirrors, m and n

with angle of incidence and thus the absorptivity and emissivity do not either. Directional variations and polarization are neglected in using the concept.

To modify the radiosity-irradiation formulation developed in Section 3, one defines a mirror-image-augmented shape factor

$$F^*_{i-j} = F_{i-j} + \Sigma_m \rho_{S,m} F_{i-j(m)}$$

$$+ \Sigma_n \Sigma_m \rho_{S,n} \rho_{S,m} F_{i-j(m,n)}$$

$$+ \Sigma_n \cdots \Sigma_m \rho_{S,n} \cdots \rho_{S,m} F_{i-j(m,\ldots,n)} \tag{4-1}$$

Only the diffuse portion of the radiosity of the mirrors appears directly in the formulation. One has

$$q^+_{D,m} = \varepsilon_m B_m + \rho_{D,m} q^-_m \tag{4-2}$$

where $\rho_{D,m}$ and ε_m no longer sum to unity, but

$$\rho_{D,m} = 1 - \varepsilon_m - \rho_{S,m} \tag{4-3}$$

The irradiation is no longer given by Eq. (3-14) but by

$$q^-_m = \sum_{j=1}^{N} F^*_{m-j} q^+_{D,j} \tag{4-4}$$

Equations (4-2) and (4-4) are, as were Eqs. (3-13) and (3-14), 2N equations in 2N unknowns, and Eq. (4-2) can be used to reduce the set to N equations in N unknowns by eliminating say $q^+_{D,j}$, leaving

$$q^-_m = \sum_{j=1}^{N} F^*_{m-j}[\varepsilon_j B_j + \rho_{D,j} q^-_j] \tag{4-5}$$

Equation (4-5) can be solved by matrix inversion or by re-iteration. In the latter case one may start with $q^-_j = 0$ and use Eq. (4-5) to find a better set of values, and repeat the process until the set of q^-_j converges.

Once Eq. (4-5) is solved, the net heat fluxes and heat flows are found from Eq. (3-18), which still holds.

$$q_m = \varepsilon_m(B_m - q^-_m) \tag{4-6}$$

$$\dot{Q}_m = \varepsilon_m A_m(B_m - q^-_m) \tag{4-7}$$

As before the transfer factor $\mathcal{F}_{i-j}$ can be found by setting $B_i = 1$ and $B_j = 0$ for $j \neq i$. Then one column of the $A_i\mathcal{F}_{i-j}$ matrix is $-Q_j$.

4.C <u>The Monte Carlo Algorithm</u>. When the engineering designer or analyst wants to consider directional variations, polarization, and other complicating factors, the Monte Carlo algorithm is perhaps the most generally applicable method, and one that is rather easily used. The Monte Carlo technique was applied to radiation heat transfer problems by Howell and coworkers in a number of papers reviewed in Ref. [7]. It is a simplified, computerized, statistical-sampling approach to ray tracing. According to electromagnetic theory, of the energy flux in an incident wave, definite fractions are reflected, absorbed, and perhaps transmitted in a wall encounter. The Monte Carlo algorithm compares a random number to the theoretical fractions and, based upon the comparison, assigns the whole incident flux to the reflected or absorbed or transmitted wave. When the calculational procedure is repeated a large number of times, the net result is correct: Of the total flux of all the rays certain fractions are reflected, absorbed, and transmitted. The heart of the Monte Carlo algorithm is to avoid branching during a ray tracing procedure. The energy is not *both* reflected and transmitted; it is *either* reflected or transmitted, and the one result is traced further. The Monte Carlo technique has the great advantage of computing the script-F transfer factor directly without any need to obtain shape factors [8,9].

The elements of a Monte Carlo program are the following:

1. Choosing a point of emission.
2. Choosing a direction of emission.
3. Tracing a ray to a wall and determining the wall node number.

4. Deciding whether the ray is absorbed, reflected or transmitted.

5. Choosing a direction of reflection or transmission.

6. Scoring increments of transfer factor.

Each will be briefly discussed.

Consider emitting surface i with $B_i = 1$ and all $B_j = 0$, $j \neq i$. If N rays are imagined to leave i, all with energy ε_i/N (where ε_i is the hemispherical emissivity of surface i), the sum of the energies of the rays absorbed by surface j is the transfer factor $\mathscr{F}_{i-j}$. If surface i has a uniform temperature and finish, it is clear than an area-weighted average should be taken. Thus points of origin on area A_i should be chosen randomly in such a way that a large number of them will be uniformly distributed over the surface. For a simple rectangle located in the domain $0 \leq x \leq a$ and $0 \leq y \leq b$ one simply chooses two random numbers between 0 and 1, P_1 and P_2, and lets the point of origin be $x = aP_1$, $y = bP_2$.

More generally area A_i might be located between $y_1(x)$ and $y_2(x)$ in the domain $x_1 \leq x \leq x_2$. Then one of two stratagems may be employed. One is hard to program but usually makes efficienct use of the computer, and the other is easy to program but often makes inefficient use of the computer. Obviously the one selected will depend upon the cost and amount of programming time versus the cost and amount of computing time needed. The computationally efficient stratagem is to compute a fractional function

$$f_A(x) = \frac{\int_{x_1}^{x} [y_2(x)-y_1(x)]dx}{\int_{x_1}^{x_2} [y_2(x)-y_1(x)]dx} \tag{4-8}$$

Then with two independent random numbers P_1 and P_2

$$x = x_1+f_A^{-1}(P_1) \quad \text{and} \quad y = y_1(x)+P_2[y_2(x)-y_1(x)] \tag{4-9}$$

The simple-to-program method is simply to choose

$$x = x_1+(x_2-x_1)P_1 \quad \text{and} \quad y = y_{min}+(y_{max}-y_{min})P_2$$

If $y_1(x) \leq y \leq y_2(x)$ then the point is used. If not, then another *pair* of P_1 and P_2 values is tried. It is essential that both P_1 and P_2 be discarded and two new random values chosen.

Next one needs to choose a direction of origin. Again two approaches are possible. A fractional function may be defined (we suppose $\varepsilon(\theta,\phi) = \varepsilon(\theta)$ to illustrate)

$$f_\varepsilon = \frac{\int_0^{\theta} \varepsilon(\theta)\cos\theta \sin\theta \, d\theta}{\int_0^{\pi/2} \varepsilon(\theta)\cos\theta \sin\theta \, d\theta}$$

Then with new random numbers P_1 and P_2

$$\theta = f_\varepsilon^{-1}(P_1) \;, \quad \phi = 2\pi P_2$$

Alternatively one can merely choose

$$\theta = \sin^{-1}\sqrt{P_1}\ , \quad \phi = 2\pi P_2$$

Here, due to the nature of black body radiation, a uniform weighting in cos θ dΩ is taken, but the ray for the direction in question is further weighted by $\varepsilon_i(\theta,\phi)/\varepsilon_i$. If polarization is being considered, the ray in addition to its weight ε_i/N or $[\varepsilon_i/N][\varepsilon(\theta,\phi)/\varepsilon_i]$ is assigned a four component vector; see Amar and Edwards [32].

With the point of origin (x_0,y_0,z_0) known and the direction fixed, the straight line ray can be represented as

$$x = x_o+Rr_x$$

$$y = y_o+Rr_y$$

$$z = z_o+Rr_z$$

where r_x, r_y, and r_z are the direction cosines; for example, if the z-axis and surface normal to i are parallel,

$$r_x = \sin\theta\cos\phi, \quad r_y = \sin\theta\sin\phi, \quad r_z = \cos\theta$$

The intersection of the ray with the plane (or other surface) of each of the other surfaces j can be found, and the intersection tested to see whether it lies within the surface limits. The intersection with the smallest positive ray distance R is the first one hit.

The angle of incidence between the ray and the normal to the wall $\underline{n}$ is found from

$$\cos\theta = -\underline{n}\cdot\underline{r} = -n_x r_x - n_y r_y - n_z r_z$$

and, when polarization is included, the new s and p directions are found using Eq. (2-93). The vector components of I are twisted into the new s and p coordinates by the rotation matrix, Eq. (2-100) with Eq. (2-97). For the angle of incidence and the state of polarization, the directional absorptivity α and reflectivity ρ are found.

Score keeping can be done in either of two ways. One procedure is to record a fraction α of the energy remaining in the ray before it struck the wall as having added to the i-j transfer factor. The remaining energy is assigned to the reflected ray. (When the reflected ray energy falls below a chosen minimum threshold all of it may be assigned to a remainder.) Another procedure is to generate a random number P. If it is less than or equal to α, the remaining energy is all absorbed. When it is greater than α, the remaining energy is all reflected.

If a further reflection is traced, the direction of the reflected ray must be found. For specular reflection Eqs. (2-111, 112, 113) apply. For perfectly diffuse reflection two new random numbers are generated, and θ with respect to the normal $\underline{n}$ is $\sin^{-1}\sqrt{P_1}$ and ϕ with respect to $\underline{x}$ is $2\pi P_2$. For imperfectly diffuse reflection θ and ϕ can be found as just described, but the weighting of the ray must be divided by the directional reflectance and multiplied by the bidirectional reflectance for the directions in question. Fractional functions could be used instead at some expense in analysis and programing time.

When the chosen number of rays (usually 4000 to 10000) is traced, the totals scored in the j transfer factor registers are normalized and printed, and another i is selected.

A refractory wall is accommodated by causing absorbed radiation to be reemitted, in the case of a total calculation with nonselective surfaces.

Consider, for example, calculating by the Monte Carlo technique the curve in Figure 3-10 for the perfectly diffuse, gray, refractory-walled, black-ended, infinitely wide slot. A sample program in Fortran is shown in Table 4-1.

4.D <u>Specular-Walled Passages</u>. Ray tracing, with or without use of the Monte Carlo algorithm, is easily executed for specular-walled passages. Figure 4-2 shows calculated results from Ref. [13] versus length-to-hydraulic-diameter ratio with parameter ρ_N, the normal reflectivity of the side wall. The hydraulic diameter for the square is the side and for the slot is twice the spacing. Full account was taken of polarization and directional variations using absorptive index k equal to refractive index n for ρ_N of 0.50 and greater and k = 0 for smaller ρ_N. At small values of L/D that parameter governs independent of passage cross-section; hence the curves for the two cross-sections come together for equal values of

TABLE 4-1

A SAMPLE MONTE-CARLO PROGRAM

```
      PRØGRAM SLØT
C     SLØT CØMPUTES SCRIPT F12 FØR A DIFFUSE REFRACTØRY SLØT
      CØMMØN IX, PI
      READ(5,11) ZL
   11 FØRMAT(F10.5)
      IX=529814367
      PI=3.1415927
      F12=0.
      DØ 101 I=1,4000
      ZØ=0.
      CALL RANDU(YØ)
      CALL DIRECT (1,RY,RZ)
  100 CØNTINUE
      IF(RY.LT.0.) GØ TØ 102
      IF(RY.EQ.0.) GØ TØ 103
      J=2
      R=(1.-YØ)/RY
      Z=ZØ+R*RZ
      ZØ=Z
      YØ=1.
  200 CØNTINUE
      IF(Z.LT.0.) GØ TØ 101
      IF(Z.GT.ZL) GØ TØ 201
      CALL DIRECT(J,RY,RZ)
      GØ TØ 100
  102 J=3
      R=-YØ/RY
      Z=ZØ+R*RZ
      ZØ=Z
      YØ=0.
      GØ TØ 200
  103 IF(RZ.LE.0.) GØ TØ 101
  201 F12=F12+1.
  101 CØNTINUE
      F12=F12/4000.
      WRITE(6,12) ZL,F12
   12 FØRMAT(2X,8HFØR ZL= ,F7.2,2X,5HF12= ,E12.5)
      STØP
      END

      SUBRØUTINE DIRECT(J,RY,RZ)
      CØMMØN IX,PI
      CALL RANPU(P)
      S=SQRT(P)
      C=SQRT(1.-P)
      CALL RANDU(P)
      PHI=2.*PI*P
      IF(J.EQ.2) GØ TØ 200
```

TABLE 4.1 (continued)

```
      IF(J.EQ.3) GØ TØ 300
      RZ=C
      RY=S*SIN(PHI)
      RETURN
  200 RY=-C
      RZ=S*CØS(PHI)
      RETURN
  300 RY=+C
      RZ=S*CØS(PHI)
      RETURN
      END

      SUBRØUTINE RANDU(P)
C     RANDU GENERATES RANDØM NUMBERS ØN AN IBM360
      CØMMØN IX,PI
      IY=IX*65539
      IF(IY) 5,6,6
    5 IY=IY+2147483847+1
    6 P=IY
      P=P*.4656613E-9
      IX=IY
      RETURN
      END
```

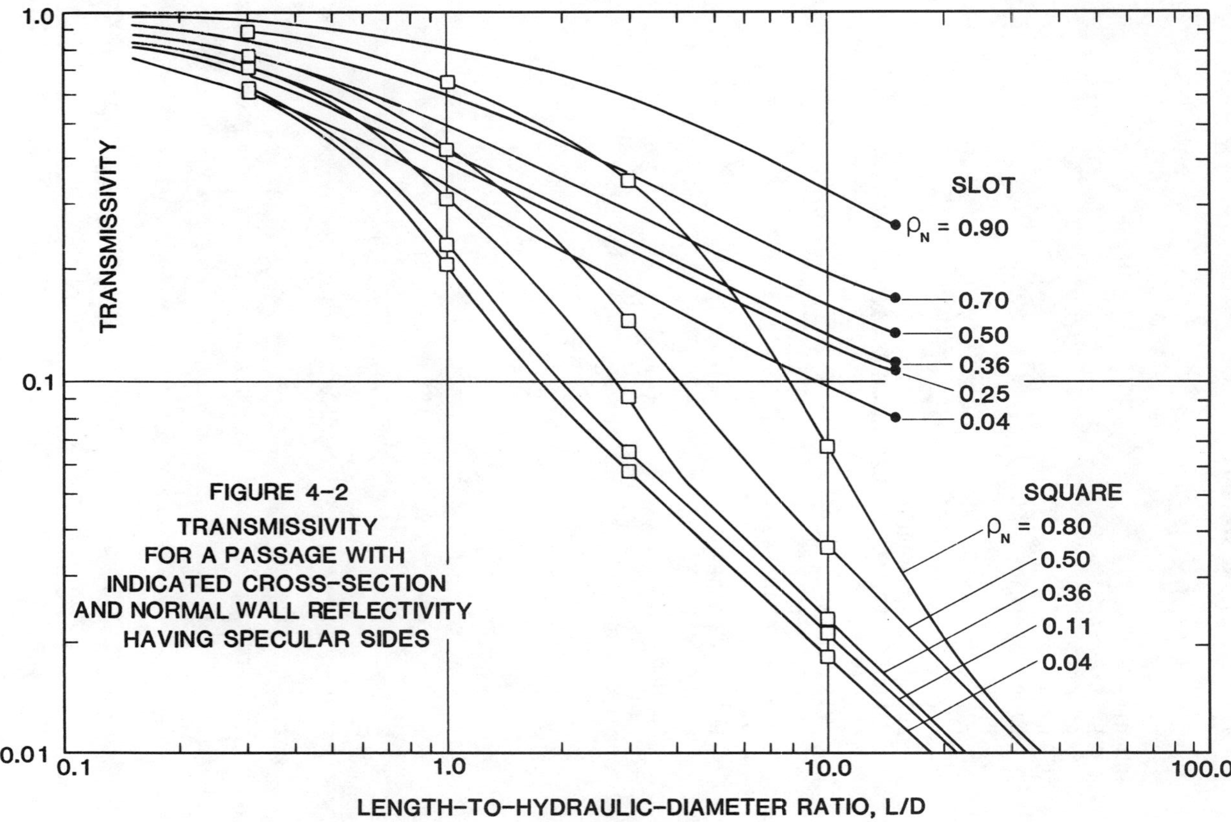

FIGURE 4–2
TRANSMISSIVITY
FOR A PASSAGE WITH
INDICATED CROSS-SECTION
AND NORMAL WALL REFLECTIVITY
HAVING SPECULAR SIDES

ρ_N. However, because of polarization in the slot and extra reflections in the square, the curves differ drastically when L/D becomes large.

4.D(a) <u>Isothermal Sides</u>. The passage transmissivity τ is directly useful when the side walls are isothermal and the ends open into hohlraums. Denoting the ends as 1 and 2 respectively and the sides as 3, one recognizes that τ is the transfer factor from 1 to 2, and $1-\tau$ is the transfer factor from 1 to 3 or 2 to 3. Hence,

$$\dot{Q}_1 = \tau A[B_1-B_2]+(1-\tau)A[B_1-B_3] \tag{4-10}$$

$$\dot{Q}_3 = (1-\tau)A[B_3-B_1+B_3-B_2] \tag{4-11}$$

where A is the cross-sectional area. Again B is used to stand for σT^4 in the case of gray walls or the Planck function $\pi I_b(\lambda,T)$ when spectral variations in properties are considered.

When the ends are nonblack, a radiosity-irradiation analysis is easily made (alternatively delta-wye, wye-delta transformations can be made).

$$q_1^+ = \varepsilon_1 B_1+(1-\varepsilon_1)q_1^- \tag{4-12a}$$

$$q_1^- = (1-\tau)B_3+\tau q_2^+ \tag{4-12b}$$

$$q_2^+ = \varepsilon_2 B_2+(1-\varepsilon_2)q_2^- \tag{4-12c}$$

$$q_2^- = (1-\tau)B_3+\tau q_1^+ \tag{4-12d}$$

Solving for q_1^-, substituting into $q_1 = \varepsilon_1 B_1-\varepsilon_1 q_1^-$, and multiplying by A gives

$$\dot{Q}_1 = A\mathscr{F}_{1-2}(B_1-B_2) + A\mathscr{F}_{1-3}(B_1-B_3) \tag{4-13a}$$

where

$$\mathscr{F}_{1-2} = \frac{\varepsilon_1\varepsilon_2\tau}{1-\tau^2(1-\varepsilon_1)(1-\varepsilon_2)} \tag{4-13b}$$

$$\mathscr{F}_{1-3} = \frac{\varepsilon_1(1-\tau)(1+\tau-\tau\varepsilon_2)}{1-\tau^2(1-\varepsilon_1)(1-\varepsilon_2)} \tag{4-13c}$$

The relations for $\mathscr{F}_{2-1}$ and $\mathscr{F}_{2-3}$ can be formed by interchanging the 1 and 2 subscripts.

When the side walls are isothermal due to good lateral conduction but are adiabatic as a whole and are gray, one sets $\dot{Q}_3$ in Eq. (4-11) equal to zero and solves for B_3. When the result is substituted into Eq. (4-10) the net heat flow from 1 into 2 is found

$$\dot{Q}_{1-2} = A[(1+\tau)/2][\sigma T_1{}^4-\sigma T_2{}^4] \tag{4-14}$$

$$(B_3 = \text{constant},\ \dot{Q}_3 = 0,\ \text{gray walls})$$

4.D(b) Nonisothermal Sides. First consider the problem for a number of isothermal segments on the sides. Circumferential variations are assumed absent, or symmetry is assumed for the slot. With black ends and specular sides no reflected radiation can reverse direction along the x axis of the passage. Thus the transfer factor from end e to side element si between x_i and x_{i+1} is simply

$$A_e \mathscr{F}_{e-si} = A_{si} \mathscr{F}_{si-e} = A[\tau(x_i - x_e) - \tau(x_{i+1} - x_e)] \qquad (4\text{-}15)$$

where either side wall properties do not vary with temperature or such variation was accounted for when $\tau(x)$ was computed. By placing imaginary end areas A_{ej} and $A_{e(j+1)}$ at locations x_j and x_{j+1} and realizing that the difference between $\mathscr{F}_{si-e(j+1)}$ and $\mathscr{F}_{si-ej}$ is the side-to-side transfer factor $\mathscr{F}_{si-sj}$ one finds

$$A_{si} \mathscr{F}_{si-sj} = A_{sj} \mathscr{F}_{sj-si} = A[\tau(x_i - x_{j+1}) + \tau(x_{i+1} - x_j) - \tau(x_{i+1} - x_{j+1}) - \tau(x_i - x_j)] \qquad (4\text{-}16)$$

With the general end-to-end, end-to-side, and side-to-side transfer factors known, any desired problem can be readily formulated and solved.

The heat flow at end 1 at x = 0 is found from Eq. (4-15)

$$\dot{Q}_1 = A\tau(L)[B_1 - B_2] + A\sum_i [\tau(x_i) - \tau(x_{i+1})][B_1 - B_i] \qquad (4\text{-}17)$$

where $x_2 = L$ is the total length of the passage. In the limit as $x_{i+1} - x_i$ goes to zero the summation passes to an integral so that

$$q_1 = \dot{Q}_1/A = \tau(L)[B_1 - B_2] + \int_0^L [-\frac{d\tau}{dx}][B_1 - B(x)]dx \qquad (4\text{-}18)$$

Integration by parts presents a form in terms of τ instead of its derivative.

$$q_1 = [B_1 - B(0)] - \tau(L)[B_2 - B(L)] - \int_0^L \tau \frac{dB}{dx} dx \qquad (4\text{-}19)$$

For example, if there is no temperature jump at the ends, and a straight-line B-gradient exists, one finds from Eq. (4-19) simply

$$q_1 = [B_1-B_2]\bar{\tau}(L) \tag{4-20}$$

where

$$\bar{\tau}(L) = \frac{1}{L}\int_0^L \tau(x)dx \tag{4-21}$$

Values of $\bar{\tau}$ may be extracted from the running integrals of τ presented in Ref. [13].

4.D(c) <u>Adiabatic Sides</u>. If the passage side walls neither conduct longitudinally nor lose or gain heat out of their backs or by convection from their surfaces and are gray, they are said to be in radiative equilibrium. The side wall black-body radiosity $B(x) = \sigma T(x)^4$ then takes a shape such that

$$dq_R(x)/dx = 0 \tag{4-22}$$

where q_R is the net heat flux in the positive x direction at any x cross-section. Generalizing Eq. (4-19) to any position x gives

$$q^*(x) = \frac{q_R(x)}{B_1-B_2} = \tau(x) + \tau(L-x)\Phi(L) - \tau(x)\Phi(0)$$
$$- \int_0^x \tau(x-x')\frac{d\Phi}{dx'}dx' - \int_x^L \tau(x'-x)\frac{d\Phi}{dx'}dx' \tag{4-23}$$

where

$$\Phi = \frac{B(x)-B_2}{B_1-B_2} \tag{4-24}$$

If a direct Monte Carlo solution is not available (recall that an adiabatic surface is modeled by requiring absorbed photons to be reemitted in the Monte Carlo algorithm) but calculations of $\tau(x)$ are available, one can employ the same approximate solution by colocation used in Section 3.F by assuming $\Phi(x) = \Phi_o-(x/L)\Delta\Phi$ and requiring $q^*(0) = q^*(L/2) = q^*(L)$. There is found

$$\Delta\Phi = \frac{1+\tau(L)-2\tau(L/2)}{1+\tau(L)-2\tau(L/2)+2\bar{\tau}(L/2)-2\bar{\tau}(L)} \tag{4-25a}$$

$$\Phi_o = (1+\Delta\Phi)/2 \tag{4-25b}$$

$$\mathscr{F}_{1-2,b} = \tau(L/2)+[\bar{\tau}(L/2)-\tau(L/2)]\Delta\Phi \tag{4-25c}$$

Figure 4-3 shows approximate values [13] found in this manner. This figure may be compared with Fig. 3-10 for diffuse-refractory-walled passages. The differences are striking. First Fig. 3-10 does not depend upon side wall reflectivity at all, because there is no difference between diffusely reflected and diffusely reemitted radiation. Figure 4-3 depends greatly upon side wall reflectivity because reflected radiation continues in some x-direction. If specular reflectivity is unity, passage transfer factor is 100 per cent regardless of length; the passage becomes a light pipe. Another difference is the sensitivity of the transfer

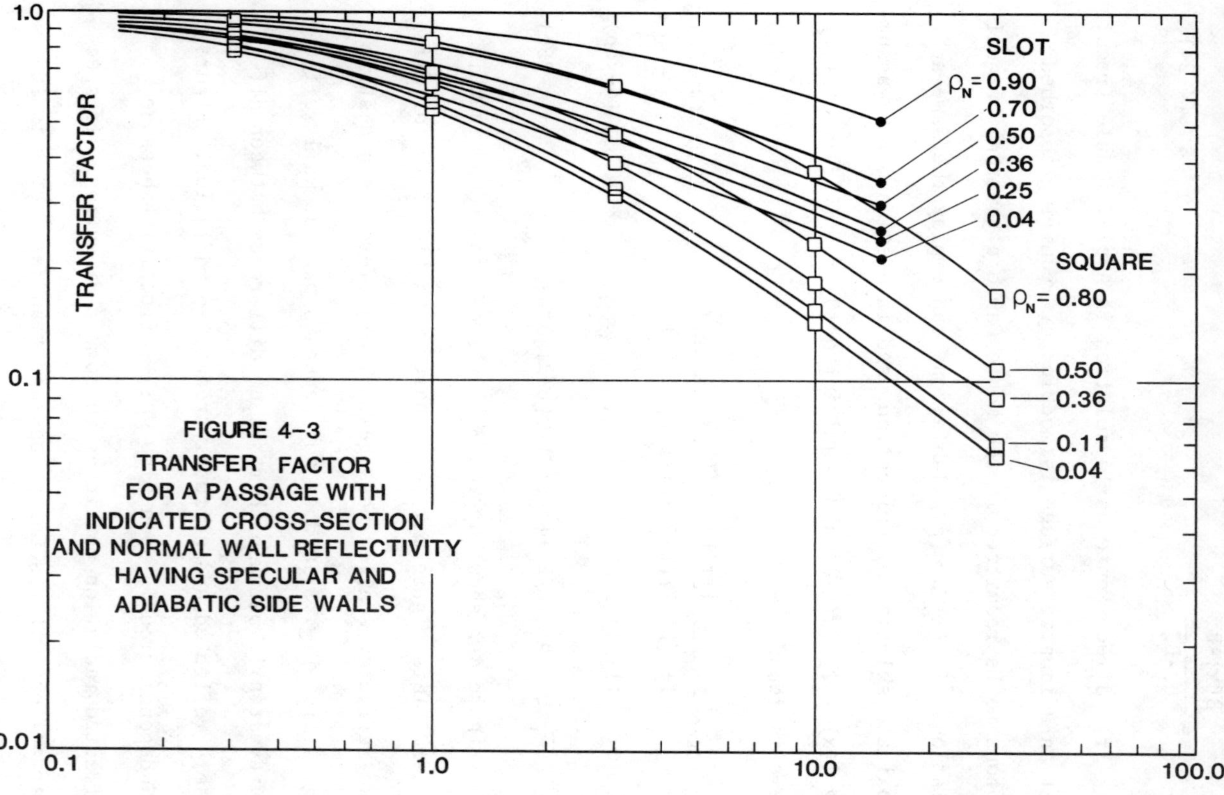

FIGURE 4–3
TRANSFER FACTOR
FOR A PASSAGE WITH
INDICATED CROSS-SECTION
AND NORMAL WALL REFLECTIVITY
HAVING SPECULAR AND
ADIABATIC SIDE WALLS

factor to the passage cross section. In Fig. 3-10 at L/D = 5 the slot transfer factor is only about 30 per cent greater than that for the circle, while in Fig. 4-3, the difference for ρ_N = 0.04 is nearly 50 per cent. The difference arises from high specular reflectivity at grazing angles and, in the slot, polarization to the s-component for which the reflectivity is still higher.

Recall that for a 3 mm gap in a 6 cm wall between a space at 600 K and 300 K a heat leak of 6.6 W/m was found at the end of Section 3.F for the diffuse side wall. For a specular side wall with 36% normal reflectivity (not atypical of an oxidized metal), Fig. 4-3 gives a transfer factor of 0.3, and the heat leak is 12.4 W/m.

4.E Surface Models. The long passage is a geometry for which the transfer factors clearly depend significantly upon the type of reflection assumed, specular or diffuse. Other geometries, e.g., the cubical specular-walled enclosure treated in Ref. [5], are not so sensitive to the type of surface reflection. The question of how to model the reflection from a technical surface thus arises mainly in concern with passage transfer factor or with similar elongated configurations.

Two surface effects need to be clearly distinguished, the effect of surface geometry and the effect of surface state. Bennett [15] suggested that differences in emissivity and absorptivity of metals associated with surface roughness were due to surface damage rather than surface geometry. Rolling [16] used platinum, which is easily annealed without surface oxidation, to show that the emissivity of rough but annealed platinum was virtually the same as that of smooth annealed platinum but quite different from that of the roughened unannealed material.

Oxidation of a metal tends to create a thin film of metal oxide dielectric or semiconductor on the surface and can be modeled via the thin film optics reviewed in Section 2.F. However, the films created by oxidizing a rough or polycrystalline metal are not necessarily even in thickness and homogeneous in depth [17].

While the surface geometry of metals appears to have a small effect on emissivity and absorptivity (apart from the change in the surface layer from mechanical damage), the surface geometry clearly affects the bidirectional reflection properties. Three types of surface models for modeling bidirectional reflectivity in radiative transfer calculations are briefly reviewed in this section.

4.E(a) Diffraction Models. For roughness elements of small height relative to the wavelength and slight slopes, Rice [18], Davies [19], and Beckmann and Spizzichino [20] have proposed models. For rms roughness σ_r and rms slope σ_r/a and an infinitely conducting metal, the Beckmann equation gives the diffuse bidirectional reflectance as

$$\rho_{B\lambda}(\theta^+,\phi^+,\theta^-,\phi^-,\lambda) = \frac{\pi^2(a/\lambda)^2}{\cos\theta^+\cos\theta^-} BFe^{-G} \tag{4-26a}$$

where

$$B = \frac{1+\cos\theta^+\cos\theta^-+\sin\theta^+\sin\theta^-\cos(\phi^+-\phi^-)}{\cos\theta^++\cos\theta^-} \tag{4-26b}$$

$$G = [2\pi(\sigma_r/\lambda)(\cos\theta^++\cos\theta^-)]^2 \tag{4-26c}$$

$$F = \sum_{m=1}^{\infty} \frac{G^m}{mm!} \exp\{-(\pi/m)(a/\lambda)[\sin^2\theta^++\sin^2\theta^- + 2\sin\theta^+\sin\theta^-\cos(\phi^+-\phi^-)]\} \tag{4-26d}$$

The specular reflectance [21,22] is

$$\rho_S(\theta^-,\lambda) = \exp\{-[4\pi\sigma_r\cos\theta^-/\lambda]^2\} \tag{4-27}$$

Houchens and Hering [22] found that the Davies relation failed to obey conservation of energy outside of a narrow range of parameters. Beckmann's result above was better in that respect, giving a useful range of $\sigma_r/\lambda = 0.01$ to 0.20 and $a/\sigma_r = 5$ to 75.

It was noted (see Ref. 2 of Section 2) that the slightly rough surfaces modeled by the Beckmann equation could be regarded from an engineering point of view as specular. The angles through which the radiation is scattered go like $\lambda/\pi a$ and $\lambda/(\pi a\cos\theta^-)$ while the fraction scattered goes like $1-\exp\{-[(4\pi\sigma_r\cos\theta^-/a)^2(a/\lambda)^2]\}$. For surfaces with σ_r on the order of 0.5 μm and slopes on the order 0.1 or less, the quantity a is 2.5 μm or more. When a/λ is large the amount of reflected radiation diffused is large, but the angle through which it is diffused is small. When a/λ is small, the angle through which the reflected radiation is diffused is large, but the amount of diffusely reflected radiation is small. For heat transfer purposes such surfaces may be modeled as specular.

4.E(b) Geometric Optics Models. Torrance and Sparrow [23] suggested a simple model to represent the near-field reflection from a surface with large roughness elements compared to the wavelength of the irradiation. A statistical distribution of facet slopes in the form $e^{-c^2\alpha^2}$ was taken. Each facet is one of a pair forming a symmetric V-groove whose azimuthal orientation is uniformly random, and the upper edges of all V grooves are in the same plane. The facets were taken to be perfectly specular to the first reflection and perfectly diffuse to interreflections. Edwards and Bertak [24] incorporated the Torrance-Sparrow model into the Monte Carlo method. Look and Love [25] identified the sharp tops of the Torrance-Sparrow model as a matter for concern and suggested rounding them off by introducing another parameter.

The Monte Carlo implementation of the Torrance-Sparrow model is as follows: Each surface is parameterized by optical constants n and k of the facets and a slope distribution angle

$\alpha_o = 1/c$. If one desires, one can fix k = n and compute the hemispherical reflectivity of the rough surface and use a measured value of that parameter to thus fix n and k for a given α_o. Reference [24] suggested using additionally a measured value of slot passage transmittance to fix α_o. When, during the course of a Monte Carlo ray tracing, a wall with fixed optical constants and roughness parameter α_o is encountered, three more random numbers are taken from the random number generator. One, call it P_1, is used to fix α by a look-up-interpolate table previously computed and stored (just as a given value of fractional function is used to fix a wavelength in Table 1-1).

$$P_1 = \frac{\int_0^{\alpha} e^{-\alpha^2/\alpha_o^2} \sin\alpha \cos\alpha \, d\alpha}{\int_0^{\pi/2} e^{-\alpha^2/\alpha_o^2} \sin\alpha . \cos\alpha . d\alpha} \tag{4-28}$$

The second fixes the azimuthal orientation of the V-groove

$$\beta = 2\pi P_2 \tag{4-29}$$

and the third fixes the point of entry into the V-groove across its top. Then Monte Carlo ray tracing is continued on a microscale, from facet to facet across the groove, until the ray is absorbed or reflected away from the plane of the V-groove upper edge. Upon reflection out of the V-groove system, ray tracing continues on the macro scale.

Amar and Edwards [32] suggested a few of the obvious permutations. The grooves need not be considered symmetric. Once α_{left} is found, another random number can generate α_{right} based some statistical distribution to simulate the grinding compound or surface finishing tool. The angle β need not be considered to be isotropic when grinding, machining, or rolling introduces a preferred azimuthal orientation of the roughness elements.

4.E(c) Scattering Bed Models. A coat of paint may be modeled as a cloud of pigment particles suspended in an atmosphere of paint resin or a ceramic as a cloud of voids separated by solid dielectric. Then methods such as those described in Section 5.B or 7.D may be used to generate a solution for the directional distribution of the radiation scattered away from the surface. The simplest model used by Giovanelli [26] is for an infinitely thick cloud in a medium of n = 1. Only a single parameter occurs, ω_s the albedo for single scatter. The bidirectional reflectance is ϕ independent and given by

$$\rho_B(\theta^+,\theta^-) = [\omega_s/4]H(\omega_s,u)H(\omega_s,v)/[u+v] \qquad (4\text{-}30)$$

where ω_s (a function of λ and T) is the albedo, and H is given by Chandraskehar [27] as shown in Table 4-2, and

$$u = \cos\theta^+ \ , \quad v = \cos\theta^- \qquad (4\text{-}31)$$

The directional reflectance is given by

TABLE 4-2

CHANDRASEKHAR'S H FUNCTION AND ITS HEMISPHERICAL AVERAGE
From Ref. [27]

ω_s	$H(\omega_s,\mu)$										$H_H(\omega_s)$
	$\mu = 0.05$	.15	.25	.35	.45	.55	.65	.75	.85	.95	
0.2	1.016	1.033	1.044	1.052	1.058	1.064	1.068	1.072	1.075	1.077	1.066
0.4	1.034	1.072	1.098	1.117	1.133	1.145	1.156	1.165	1.173	1.180	1.152
0.6	1.055	1.120	1.167	1.203	1.233	1.258	1.279	1.298	1.314	1.329	1.273
0.8	1.082	1.187	1.266	1.332	1.388	1.437	1.480	1.518	1.553	1.584	1.472
0.9	1.100	1.235	1.343	1.436	1.518	1.592	1.658	1.719	1.775	1.826	1.651
0.95	1.112	1.269	1.401	1.517	1.622	1.719	1.809	1.892	1.970	2.042	1.804
0.975	1.120	1.294	1.443	1.578	1.703	1.820	1.930	2.033	2.132	2.226	1.929
1.000	1.137	1.351	1.547	1.736	1.921	2.104	2.284	2.464	2.642	2.819	2.309

$$\rho(\theta^-) = 1-[1-\omega_s]^{1/2}H(\omega_s,v) \tag{4-32}$$

The hemispherical reflectance is then

$$\rho_H(\theta^-) = 1-[1-\omega_s]^{1/2}H_H(\omega_s) \tag{4-33}$$

where the subscript denotes a hemispherical average as in Eq. (2-4).

$$H_H(\omega_s) = \int_0^1 H(\omega_s,v)\,2v\,dv \tag{4-34}$$

When the scattering medium is finite (as for a thin coat of paint) additional parameters are the optical constants of the substrate and the optical depth of the bed as well as its albedo for single scatter. If the paint resin has a refractive index different from unity that becomes another parameter. Finally the resin surface can be rough [29]. Clearly a glossy enamel has a smooth resin completely submerging the pigment particles while a flat or matt paint does not.

4.F Rough-Walled Passages. Table 4-3 shows results [32] calculated using the Torrance-Sparrow model and taking full account of polarization. For the passage cross-sections and lengths, whether one included polarization or not made essentially no difference for the rough walls and less than 5% difference for the smooth walls. Using a fixed value equal to the normal reflectivity in place of a directionally varying reflectivity gave rise to 10% errors. Using a fixed value of facet reflectivity equal to the hemispherical reflectivity

TABLE 4-3

ROUGH, ISOTHERMAL-WALLED PASSAGE TRANSMISSIVITY AND ABSORPTIVITY
(Facet optical constants: $n = k = 50$ giving $\rho_H = 0.95$)
From Ref. [32]

Roughness Parameter α_o Degrees	Cross Section*	Passage Transmissivity $\mathcal{F}_{1-2}$				Passage Absorptivity $\mathcal{F}_{1-3}$			
		$\frac{L}{D}$ or $\frac{L}{H} = 1$	2	4	8	$\frac{L}{D}$ or $\frac{L}{H} = 1$	2	4	8
0	Circle	0.922	0.860	0.760	0.620	0.078	0.140	0.240	0.380
	Square	0.921	0.862	0.761	0.624	0.079	0.138	0.239	0.376
	1x2	0.941	0.891	0.810	0.683	0.059	0.109	0.190	0.317
10	Circle	0.851	0.752	0.594	0.390	0.090	0.168	0.297	0.483
	Square	0.847	0.736	0.581	0.374	0.093	0.180	0.305	0.499
	1x2	0.880	0.780	0.653	0.457	0.069	0.138	0.245	0.416
20	Circle	0.745	0.588	0.399	0.208	0.098	0.186	0.329	0.479
	Square	0.740	0.589	0.385	0.170	0.101	0.195	0.334	0.518
	1x2	0.792	0.652	0.477	0.251	0.076	0.151	0.273	0.448
40	Circle	0.626	0.423	0.227	0.083	0.117	0.222	0.362	0.437
	Square	0.622	0.436	0.228	0.068	0.120	0.217	0.351	0.494
	1x2	0.681	0.511	0.309	0.122	0.091	0.176	0.302	0.440

*Cross-sections are a circle of diameter D, a square of side D, and a rectangle of 2H by H. The hydraulic diameters are respectively D, D, and 4H/3.

was adequate for the rough surfaces but for the smooth ones caused errors up to 16 per cent. Since the slot was not among the cross-sections investigated, the results correlate quite well with the length-to-hydraulic-diameter ratio for a given value of α_o. Interpreting calculated passage transmissivities using the Torrance and Sparrow model to indicate specularity according to the specular diffuse model (wherein X_S of the reflection is specular and $1-X_S$ is diffuse) gave the following values [24]

α_o	X_S
0°	1.00
10°	0.94
20°	0.77
40°	0.59

The results suggest that even very rough surfaces behave for heat transfer purposes in a significantly specular manner. However, examination of Table 4-3 shows that passage transmittance is quite sensitive to α_o so that complete specularity ($\alpha_o = 0$) cannot be assumed for very rough surfaces. The idea of using a measured passage transmissivity to fix α_o or X_S can be made the basis of a specularity or roughness-measuring instrument.

REFERENCES TO SECTION 4

I. The Mirror-Image Concept

1. Eckert, E. R. G., and E. M. Sparrow, "Radiative Heat Exchange Between Surfaces with Specular Reflection", International J. Heat and Mass Transfer, Vol. 3, pp. 42-54, 1961.

2. Sparrow, E. M., E. R. G. Eckert, and V. K. Jonsson, "An Enclosure Theory for Radiative Exchange Between Specularly and Diffusely Reflecting Surfaces", J. Heat Transfer, Vol. 84, pp. 294-300, 1964.

3. Bobco, R. P., "Radiation Heat Transfer in Semigray Enclosures with Specularly and Diffusely Reflecting Surfaces", J. Heat Transfer, Vol. 86, pp. 123-130, 1964.

4. Sparrow, E. M., and S. L. Lin, "Radiation Heat Transfer at a Surface Having Both Specular and Diffuse Reflectance Components", International J. Heat and Mass Transfer, Vol. 8, pp. 769-779, 1965.

5. Sarofim, A. F., and H. C. Hottel, "Radiative Interchange Among Non-Lambert Surfaces", J. Heat Transfer, Vol. 88, pp. 37-44, 1966.

6. O'Brien, P. F., and R. P. Bobco, "Interreflections in Mirrored Rooms", Illuminating Engineering, Vol. 59, pp. 337-344, 1964.

II. The Monte Carlo Algorithm

7. Howell, J. R., "Application of Monte Carlo to Heat Transfer Problems", Advances in Heat Transfer, Vol. 5, pp. 1-54, 1966.

8. Weiner, M. M., J. W. Tindall, and L. M. Candell, "Radiative Interchange Factors by Monte Carlo", ASME Paper 65-WA/HT-51 (1965).

9. Corlett, R. C., "Direct Monte Carlo Calculation of Radiative Heat Transfer in Vacuum", J. Heat Transfer, Vol. 88, pp. 376-382, 1966.

10. Toor, J. S., and R. Viskanta, "A Numerical Experiment of Radiant Heat Exchange by the Monte Carlo Method", International J. Heat and Mass Transfer, Vol. 11, pp. 883-897, 1968.

III. Smooth-Walled Passages (see also Fig. 5 of Ref. 4)

11. Perlmutter, M., and R. Siegel, "Effect of Specularly Reflecting Gray Surface on Thermal Radiation Through a Tube and from Its Heat Wall", J. Heat Transfer, Vol. 85, pp. 55-62, 1963.

12. O'Brien, P. F., and E. F. Sowell, "Radiant Transfer through Specular Tubes", J. Optical Soc. Am., Vol. 57, pp. 28-34, 1967.

13. Edwards, D. K., and R. D. Tobin, "Effect of Polarization on Radiant Heat Transfer through Long Passages", J. Heat Transfer, Vol. 89, pp. 132-138, 1967.

14. Felland, J. R., and D. K. Edwards, "Solar and Infrared Radiation Properties of Parallel-Plate Honeycomb", AIAA J. of Energy, Vol. 2, pp. 309-317, 1978.

IV. Surface Models

A. Surface Damage

15. Bennett, H. E., "Influence of Surface Roughness, Surface Damage, and Oxide Films on Emittance", NASA SP-55, 1965, pp. 145-152.

16. Rolling, R., "Effect of Surface Roughness on the Spectral and Total Emittance of Platinum", Progress in Aeronautics and Astronautics, Vol. 20, pp. 91-114, 1967.

17. Edwards, D. K., and I. Catton, "Radiation Characteristics of Rough and Oxidized Metals", Advances in Thermophysical Properties at Extreme Temperatures and Pressures, Am. Soc. Mech. Engrs., New York, 1965, pp. 189-199.

B. Diffraction Theory

18. Rice. S. O., "Reflection of Electromagnetic Waves from Slightly Rough Surfaces", Communications on Pure and Applied Mathematics, Vol. 4, pp. 351-378, 1951.

19. Davies, H., "The Reflection of Electromagnetic Waves from a Rough Surface", Proceedings of the Institution of Electrical Engineers, Part IV, Vol. 101, pp. 209-214, 1954.

20. Beckmann, P., and A. Spizzichino, The Scattering of Electromagnetic Waves from Rough Surfaces, Macmillan, New York, 1963.

21. Bennett, H. E., and J. O. Porteus, "Relation Between Surface Roughness and Specular Reflectance at Normal Incidence", J. Optical Soc. Am., Vol. 51, pp. 123-129, 1961.

22. Houchens, A. F., and R. G. Hering, "Bidirectional Reflectance of Rough Metal Surfaces", Progress in Astronautics and Aeronautics, Vol. 20, pp. 65-89, 1967.

C. Geometric Optics

23. Torrance, K. E., and E. M. Sparrow, "Theory for Off-Specular Reflection from Roughened Surfaces", J. Optical Soc. Am., Vol. 57, pp. 1105-1114, 1967.

24. Edwards, D. K., and I. V. Bertak, "Imperfect Reflections in Thermal Radiation Transfer", Progress in Astronautics and Aeronautics, Vol. 24, pp. 143-163, 1970.

25. Look, D. C., and T. J. Love, "Investigation of the Effects of Surface Roughness upon Reflectance", Progress in Astronautics and Aeronautics, Vol. 24, pp. 123-142, 1970.

D. Scattering Beds

26. Giovanelli, R. G., "Reflection by Semi-Infinite Diffusers", Optica Acta, Vol. 2, pp. 153-162, 1955.

27. Chandrasekhar, S., Radiative Transfer, Dover, New York, 1960, pp. 107, 328.

28. Hottel, H. C., and A. F. Sarofim, Radiative Transfer, McGraw-Hill, New York, 1967, pp. 417-418.

29. Rogers, J. E., and D. K. Edwards, "Bidirectional Reflectance and Transmittance of a Scattering-Absorbing Medium with a Rough Surface", Progress in Astronautics and Aeronautics, Vol. 49, pp. 3-24, 1976.

30. Armaly, B. F., and H. S. El-Baz, "Influence of Substrate Properties on the Apparent Emittance of an Isothermal Isotropically Scattering Medium", J. Heat Transfer, Vol. 99, pp. 208-211, 1977.

31. Crosbie, A. L., "Apparent Radiative Properties of an Isotropically Scattering Medium on a Diffuse Substrate", J. Heat Transfer, Vol. 101, pp. 68-75, 1979.

V. Rough-Walled Passages (see also Refs. [4,24])

32. Amar, R. C., and D. K. Edwards, "Reflection and Transmission by Rough-Walled Passages", Progress in Astronautics and Aeronautics, Vol. 31, pp. 475-495, 1973.

EXERCISES FOR SECTION 4

1. Consider a four-sided enclosure formed by a long square cross-section. Source 1 is opposite sink 3, and second source 2 is below refractory R. Surfaces 1, 2, and 3 are black. Find transfer factors $\mathscr{F}_{1-2}$, $\mathscr{F}_{1-3}$, and $\mathscr{F}_{2-3}$ including the effect of the refractory where the refractory is (a) perfectly diffuse and (b) perfectly specular with $\rho = 1$.

2. An instrument bay is in the form of a long square cross-section. Diffuse source 1 ($\varepsilon_1 = 0.85$) is the top. Diffuse sink 2 ($\varepsilon_2 = 0.85$) is an adjacent side. Opposite the source is a diffuse refractory 3. The other side 4 adjacent to the source is also a refractory but is not necessarily diffuse. Find the transfer factor from 1 to 2 including the effect of the refractories when (a) side 4 is diffuse, (b) side 4 is prefectly specular with $\rho = 1$. (c) Find the power transfer in W/m^2 when the cross-section is 10 cm by 10 cm, and $T_1 = 60°C$, and $T_2 = 27°C$.

3. Find the heat loss per meter of width through a slot 5 mm high, and 5 cm long between a space at 1000 K and one at 300 K when (a) both walls are diffuse refractories and (b) one wall is a diffuse refractory and the other wall is a specular mirror with $\rho = 1$.

4. Using in sequence the quasi-random numbers in Table A, illustrate your answers to the following questions on the Monte Carlo technique for calculation of radiative transfer. First derive the appropriate expression and then give a numerical example.

a) Find the location (x,y) from which a ray or photon bundle originates. The surface is a circle of radius 1 lying in the x-y plane with center at 0,0.

b) Find the x,y, and z components of the unit vector giving the direction of a ray emitted by a surface in the x-y plane whose directional emissivity is given approximately by $\varepsilon = 0.05\ (\cos\theta + \sec\theta)$.

c) Of 10,000 rays emitted by surface 1 with directional emissivity $\varepsilon_1 = 0.05\ (\cos\theta + \sec\theta)$, 5992 were found to have been absorbed by surface 2. What is the script-F transfer factor from 1 to 2 according to your Monte Carlo Algorithm?

d) A ray strikes an opaque surface lying in the plane z = 2 and facing down. The surface emissivity is given approximately by $\varepsilon = 0.05\ (\cos\theta + \sec\theta)$, and the incident ray has direction cosines of $r_x = 0.5$, $r_y = 0.5$, and $r_z = 0.75$. Is the ray scored as absorbed or is it reflected?

e) A ray is found to be reflected by a perfectly diffuse surface whose outward normal has the following

4. (continued)

x, y, and z components: $n_x = 1$, $n_y = 0$, $n_z = 0$. The incident ray came in with unit vector $r_x = -0.5$, $r_y = -0.5$, and $r_z = +0.75$. What are the unit fector components of the reflected ray?

TABLE A. Quasi-Random Numbers

0.5228	0.2506	0.0360
0.3320	0.0036	0.6096
0.0224	0.6493	0.1612
0.0822	0.1590	0.5985
0.4717	0.3281	0.8202
0.2501	0.7650	0.2728
0.2550	0.5225	0.4420
0.5025	0.3006	0.5364

5. Write a Monte Carlo program for the situation in Exercise 3-10. Find the solar irradiation upon each window.

6. The transmission through a particular specular-walled passage varies approximately as $(1+0.27x)^{-2}$ where x is L/D_H. What would be the emission out of the cold end of such a passage when $L/D_H = 10$ and the side-wall black body radiosity is linear from a value of 1 at the hot end to zero at the cold end, and no temperature jump exists between the side wall and black ends?

7. Find the hemispherical reflectivity of a paint whose pigment particles have an albedo for single scatter of (a) 0.975, (b) 0.95, (c) 0.90, and (d) 0.20.

8. Find the bidirectional reflectivity of a paint for near normal incidence $\cos\theta^{-} = 0.95$ and angles of emergence given by $\cos\theta^{+} = 0.05, 0.15, 0.25, 0.35$, and 0.45, for an albedo of 0.975. Discuss what the effect of such behavior would have on the performance of the integrating sphere in Fig. 2-2.

9. The reflection from a sandblasted aluminum surface is to be modeled by assuming that a fraction F of the reflection is perfectly diffuse and the remainder is perfectly specular. To decide what value F should have, a square passage is formed with L/d = 4, and the transmissivity through it is measured to be 0.30. Estimate the value of F you would assign.

10. Write a Monte Carlo program that would allow you to answer Exercise 9 with more confidence.

5 GAS RADIATION PROPERTIES

5.A The Equation of Transfer. Up to this point, a diathermanous medium has been assumed. In such a medium the radiant intensity I of a stream of photons is unaffected by passage through the medium. More generally the photons and matter interact. Three processes may be distinguished: (1) net absorption (total absorption minus induced emission), (2) spontaneous emission, and (3) scattering. The latter can be broken down into scattering out of the beam and scattering into the beam. The result is that in slant path increment ds there is an incremental change in intensity dI as pictured in Fig. 5-1. The equation giving dI/ds is named the equation of transfer.

The absorption, emission, and scattering properties of matter are sometimes characterized by cross-sections. For example, visualize a spherical oil droplet such as is sprayed into a combustion chamber. The droplet of radius R has a total surface area of $4\pi R^2$ but its projected area is πR^2. The latter is said to be the geometric cross-section. If one examines the shadow behind a droplet that is large compared to the wavelength, one finds, due to diffraction, a shadow area of $2\pi R^2$, but a bright halo contains half of the radiant power missing due to the shadow, half of $I d\Omega 2\pi R^2$. Whether the droplet is large or small, the radiant power missing from the incident beam divided by that incident upon an area of

πR^2 is termed the extinction efficiency Q_e. Thus if one considers the halo radiation as scattered, i.e., caused to deviate in direction, the extinction efficiency of a large particle is 2; but, if one considers the halo as undeviated, a more reasonable view for engineering power transfer calculations, then Q_e is 1 for a large particle.

The power missing from the shadow may have been absorbed, or it may have been scattered into other directions. The fraction scattered into other directions is called the albedo for single scatter ω_s. The fraction $1-\omega_s$ is sometimes called the particle emissivity. Actually it would be better called the particle absorptivity, but Kirchhoff's law is invoked. The quantity $\omega_s Q_e$ is called the scattering efficiency Q_s, and $(1-\omega_s)Q_e$ is called the absorption efficiency Q_a.

Consider N_V particles per unit volume in a gas. Within volume Ads there are N_VAds particles, each casting a shadow of $Q_e \pi R^2$. The length ds is so short that none of the shadows overlap. Thus the fraction of the radiant power taken out of a beam of cross-section A by the particles is the ratio of the shadow area to the total area

$$k_e ds = \frac{N_V A ds Q_e \pi R^2}{A} = Q_e \pi R^2 N_V ds$$

The quantity k_e $[m^{-1}]$ is called the extinction coefficient

$$k_e = Q_e \pi R^2 N_V \tag{5-1}$$

The absorption and scattering coefficients follow

$$k_a = (1-\omega_s) Q_e \pi R^2 N_V = Q_a \pi R^2 N_V \tag{5-2}$$

$$k_s = \omega_s Q_r \pi R^2 N_V = Q_s \pi R^2 N_V \tag{5-3}$$

At times one prefers to think of the radiation-matter interaction in terms of the mass of matter. The density of the absorbing mass present in a sphere is

$$\rho_a = \frac{4}{3} \pi R^3 \rho_{solid} N_V$$

$$\pi R^2 N_V = \frac{3}{4R\rho_{solid}} \rho_a \tag{5-4}$$

Thus Eqs. (5-1 to 5-3) can be written (whether the matter is in spherical form or not)

$$k_e = \kappa_e \rho_a \;, \quad k_a = \kappa_a \rho_a \;, \quad k_s = \kappa_s \rho_a \tag{5-5}$$

where κ_e is the mass extinction coefficient, κ_a the mass absorption coefficient, and κ_s the mass scattering coefficient.

Although the concepts are presented here for clarity in terms of a cloud of fuel droplets, they are general and are applied equally well (with suitable Q's) to a cloud of gas molecules or soot particles, or other absorbing species.

The first term of the equation of transfer is seen to be simply $-k_e I = -(k_a + k_s) I$, the amount intensity I is reduced by absorption plus scattering. The emission term is found via Kirchhoff's law for the case of local thermodynamic equilibrium. If the thermodynamic state of the particle or molecule can be characterized by a temperature T, then the emission is $k_a I_b(\nu,T)$. At complete thermodynamic equilibrium when $I = I_b$, the loss of $-k_a I$ is offset exactly by the gain $+k_a I_b$. The equation of transfer for a nonscattering medium in local thermodynamic equilibrium is thus simply

$$\frac{dI}{ds} = -k_a I + k_a I_b \quad (k_s=0) \tag{5-6}$$

Isotropic scattering is a mathematical concept akin to perfectly diffuse reflection. When the scattering is isotropic the intensity of the radiation scattered by a single particle is constant regardless of direction. Consider a large opaque specular sphere with reflectivity equal to ω_s, a constant independent of angle of incidence. Consider Figure 5.2. Radiation striking the sphere at angle θ_s from its nose is turned through angle θ equal to $2\theta_s$. The power intercepted is $Id\Omega \cos\theta \, dA = I_o d\Omega_o \cos\theta_s \sin\theta_s R d\phi_s R d\theta_s$, and a fraction ω_s of this power is scattered in solid angle $d\Omega = \sin\theta \, d\theta d\phi = \sin 2\theta_s \, 2d\theta_s d\phi_s$. The intensity of the scattered radiation is the scattered power per unit projected area and per unit solid angle

$$I_{scattered} = \frac{\omega_s I_o d\Omega_o \cos\theta_s \sin\theta_s R d\phi_s R d\theta_s}{\pi R^2 \sin 2\theta_s \, 2d\theta_s d\phi_s} = \omega_s I_o \frac{d\Omega_o}{4\pi} \tag{5-7}$$

FIGURE 5-1

Change of Intensity I Along Incremental Path Length ds

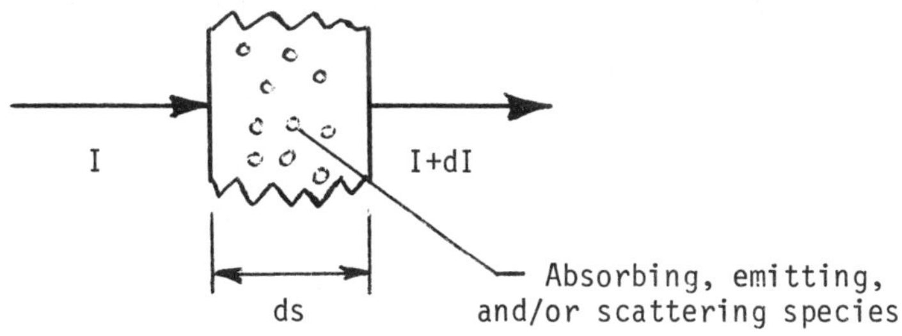

FIGURE 5-2

Scattering from a Large Opaque Specular Sphere

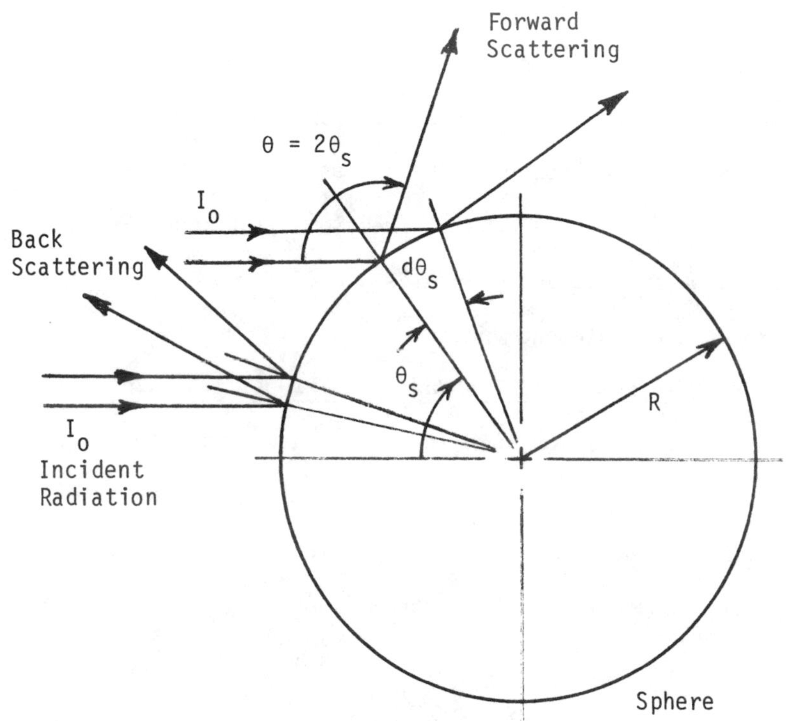

The value is found to be independent of angle θ_s or θ. For a cloud of particles of thickness ds in a beam of cross-sectional area A, the scattering into the beam is the product of fraction of the area A taken up by particles $k_e ds$ times $I_o d\Omega_0/4\pi$. Recognizing that $\omega_s k_e$ is k_s and integrating over all solid angles of incidence gives the scattering-in source term needed to complete the equation of transfer for isotropic scattering,

$$\frac{dI}{ds} = -(k_a+k_s)I + k_a I_b + \frac{k_s}{4\pi}\int_0^{4\pi} I d\Omega \qquad (5\text{-}8)$$

If the scattering is anisotropic, for example, if in the case of the large opaque specular particle the reflectivity ω_s is a function of angle θ_s, the variation with angle is accounted for by a "phase-function" p under the integral. For spherical particles, p is a function merely of θ ($= 2\theta_s$). In general, for non-spherical particles lined up by say a magnetic field, p would be a bidirectional function. When polarization is included p becomes a bidirectional 4 x 4 matrix and I a 4-element vector.

5.B <u>Measurement of Gas Radiation Properties</u>. The execucution of a calculation in which an engineer can have confidence is often difficult, because property values are often unknown. In practical situations one is often uncertain about even the composition of the medium. For example, in a burner flame or rocket motor plume, how much soot is present? Even when one thinks he knows, for example, how much ash is present in a pulverized-coal-fired chamber, one may be uncertain as to the size distribution. Ancient-Greek-type speculation, that is, calculations in the absence of measurements, does little to supply reliable design figures. What must nearly always be done is to measure the properties needed or otherwise adjust the calculational model to physical reality.

The basic experimental technique used in measuring gas radiation properties is sighting with a radiometer through a length of gas into a cavity or other reference target. The radiometer may be a total device such as a thermopile or thermistor-bridge radiometer, a low-resolution device such as a prism or variable filter spectrometer, or a high resolution spectral device such as a precision grating spectrometer or interferometer. The gas may be contained in a cell having windows or uncontained in an open jet. The windows in turn may be either hot or cold. A hybrid gas containment is a cell with open windows. Reviews of experimental measurements are given by Hottel and Sarofim [4] and Edwards [5].

Figure 5-3 from [5] shows an experimental arrangement. A low resolution detector system at the bottom of the figure

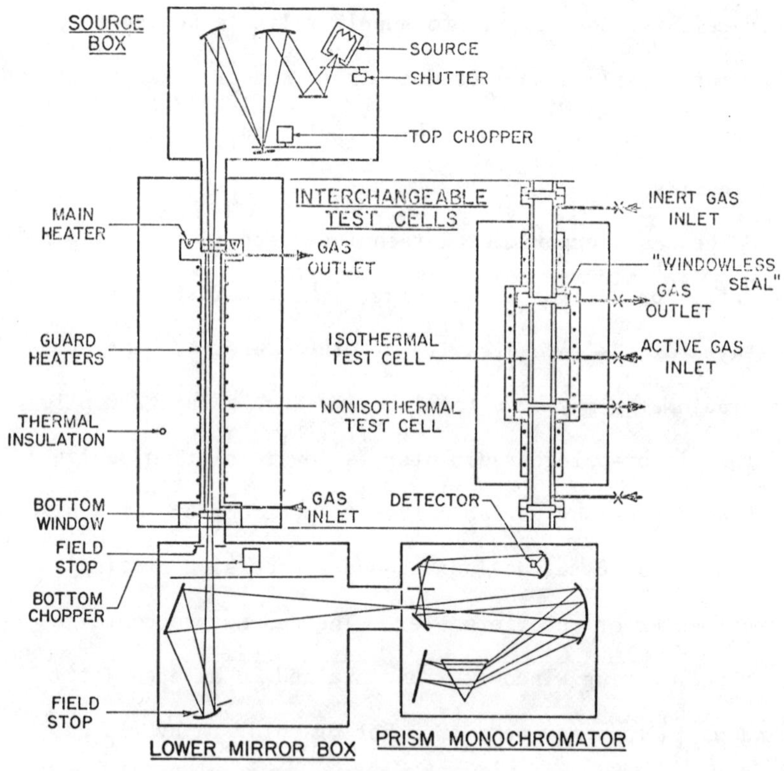

FIGURE 5-3

Gas Radiation Apparatus
from Ref. [5]

is formed by an alkaline halide (NaCl, KBr) salt crystal prism which disperses energy entering the entrance slit into an "infrared rainbow". The exit slit picks off the infrared wavelength desired for detection by a thermocouple. The radiation onto the thermocouple is composed of steady heat radiation from the hohlraum formed by the monochromator housing and a fluctuating portion coming through the exit slit. The fluctuation is caused by "chopping", interrupting the beam 13 times per second with a rotating blade. When it is desired to measure the transmissivity of the gas the source radiation is chopped at the top of the gas cell so that emission by the gas arrives upon the detector steadily. The steady irradiation of the detector gives rise to a steady heating which causes a D.C. signal or "bias". The fluctuating radiation gives rise to an A.C. signal from the detector, and it is the A.C. component that is amplified and recorded. Thus when the top chopper is used, and the bottom one is off and open, only the transmitted portion of the incident radiation is detected. When the signal from transmitted radiation is divided by the signal obtained with an evacuated cell, the transmissivity if found. Subtraction of the transmissivity from unity gives the experimentally measured absorptivity shown in Fig. 5-4.

When it is desired to measure the emission from the gas, the source at the top of the figure is made cold, the top

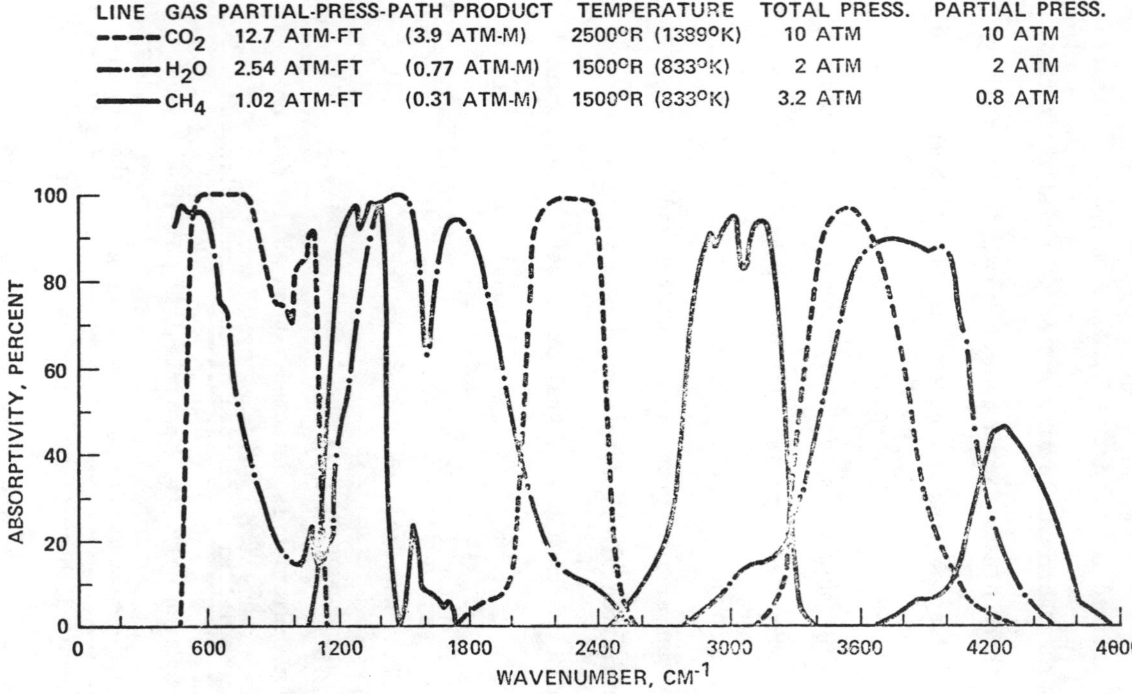

FIGURE 5-4

Low-Resolution Spectra
from Ref. [19]

chopper is off and open, and the bottom chopper is used. Thus the signal received is $(1-e^{-\kappa L})I_b(T_g,\nu)$ when the gas is isothermal. To avoid having the cell windows (also alkaline halide salt crystals) attacked by high temperature or corrosive gases, the isothermal test cell pictured in the figure was used. A heavy inert and transparent gas supported the test gas which in turn was capped by a light inert and transparent gas (helium).

Experimentally it is found that monatomic gases such as He, Ne, and Ar absorb only at wavelengths considerably shorter than 1 μm and then in lines such as those found in the solar spectrum. Much the same is true of symmetric diatomic gases such as N_2 and O_2, except that at very high pressures a "pressure-induced dipole" is created from molecular collisions. Asymmetric diatomic gases such as CO and NO and polyatomic gases such as CO_2 and H_2O are observed to absorb strongly in certain wavenumber (or wavelength) regions called "absorption bands", as illustrated in Fig. 5-4. The figure shows, for example that CO_2 has bands at 15 μm, 4.3 μm, and 2.7 μm. At temperatures well above 300 K the 9.4 and 10.4 μm bands apparent in the figures in the vicinity of 1000 cm^{-1} (recall $\nu[cm^{-1}] = 10^4/\lambda[\mu m]$) are strongly absorbing, but not at 300 K. Such bands are called "hot bands". At high pressure, say P > 5 atm, a 7.5 μm pressure-induced band appears for CO_2.

5.C <u>Qualitative Remarks on the Physics of Gas Radiation</u>. The quantum mechanical explanation of these observations is that the vibrational energy levels of the molecules are quantized, and only photons with spectrally tuned energies corresponding to the difference in energy levels can interact with the molecules. The interaction can occur only when the mode of vibration has an associated fluctuating electric dipole; hence the lack of absorption by symmetric diatomic molecules. Under high resolution the bands are seen to possess lines, that is, the vibrational energy levels are "split". The splitting is due to quantized rotational energy levels, and during absorption or emission vibrational and rotational transitions occur simultaneously. The lines become less intense but broader at high pressures, because the rotational energy levels are perturbed by molecular collisions. If the pressure is high enough the lines may overlap and not be observable. When pressure increases have an observable effect, the low resolution absorptivity in Fig. 5-4 will increase somewhat. More readily apparent is the effect of temperature. As temperature is increased the bands become wider and less intense at their center. Under high resolution, previously very weak lines are seen to grow with increasing temperature, and the previously strong lines become less intensely absorbing but narrower, the latter effect due to a decreased collision frequency, unless temperatures are so high that Doppler broadening becomes evident from frequency shifts brought about by molecular velocities toward and away from the observer. If weak lines grow between the strong ones, the lines may become overlapped even though the individual collision-broadened lines are narrower.

Temperatures considerably higher than those encountered in furnaces and combustion chambers occur in arcs, in shock-heated gases ahead of hypersonic vehicles such as planetary probes and returning space vehicles, and in nuclear blasts. At these high temperatures electronic energy level transitions, "bound-bound" transitions, give rise to lines for monatomic gas and electronic band systems for polyatomic gases. Photoionization, "bound-free" transitions occur when the photons and thermal excitation are sufficient to ionize the gas species. These transitions give rise to continuum absorption as opposed to line or band absorption because a photon may possess any energy beyond the minimum required for ionization and still interact with the atom or molecule. When the temperatures are high enough to ionize the gases significantly, colliding electrons may emit or absorb photons in so-called "free-free transitions" or "brehmstrahlung".

The minimum energy requirement for photoionization gives rise to an absorption edge such as that evident in

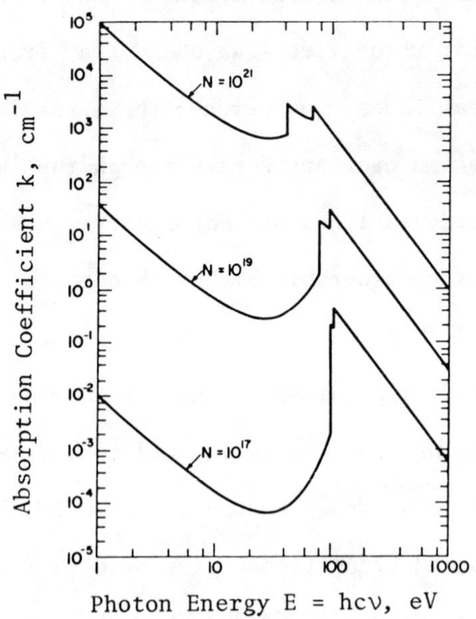

FIGURE 5-5

High-Energy Absorption Spectrum of Dissociated Nitrogen
N = Atoms/cm^3, T_g = 10 eV from Ref. [9]

Fig. 5-5 from Chapter 1 of [9] for 10^{17} atoms/cm^3 of nitrogen at a temperature of 10 eV (the electron volt, a unit of energy, is used both to designate temperature through $E = kT$, 1 eV = 11600 K, and wavenumber through $E = hc\nu$, 1 eV = 8066 cm^{-1}). When a gas mixture contains a number of species with different ionization energies, the absorption coefficient log-log plot versus wavenumber often resembles a saw-tooth curve. Notice that the edge is not entirely sharp due to the thermal distribution of energy states in the high temperature gas.

5.D Spectral, Band, and Total Property Definitions. The ideas of spectral and total properties present no problems for continuum radiation; they are the same as discussed previously for the case of surfaces. For example, a cloud containing a distribution of particle sizes gives rise to continuum radiation. By "continuum" is meant the fact that k_a and k_s and hence I vary slowly and smoothly with wavelength or wavenumber. For example, the spectral mass absorption coefficient for soot may be reasonably approximated as

$$\kappa_a = \kappa_o(\lambda/\lambda_o)^{-n} \tag{5-9}$$

where n is on the order of 1. The spectral emissivity of a beam L long in an isothermal cloud follows from integrating the equation of transfer, Eq. (5-6), with I_b constant.

Introducing Eq. (5-5) and taking I = 0 at s = 0 gives

$$I = I_b(1-e^{-\rho_a\kappa_a L}) \tag{5-10}$$

One sees that the spectral emissivity (and absorptivity) is

$$\varepsilon_\lambda = (1-e^{-\rho_a\kappa_a L}) \tag{5-11}$$

The total emissivity is then

$$\varepsilon = \frac{\int_0^\infty I d\lambda}{\int_0^\infty I_b d\lambda} = \int_0^1 [1-e^{-\rho_a\kappa_a L}]df_e(\lambda T) \tag{5-12}$$

where f_e is the external fractional function. With slow and smooth spectral variation we have found (e.g., Eq. (2-84)) that the total property can be approximated by the spectral property at $f_e(\lambda T) \doteq 0.49$. Hence a rough approximation (±7 per cent) to Eq. (5-12) for soot is

$$\varepsilon \doteq 1-e^{-\rho_a L\kappa_o(\lambda_m/\lambda_o)^{-n}} \tag{5-13a}$$

$$\lambda_m T = 4050\ \mu mK, \quad \kappa_o = 5600\ m^2/kg\ , \quad n = 1.086$$

A very good approximation (±1 per cent) to Eq. (5-12) is obtained using 3-point Gaussian quadrature

$$\varepsilon \doteq \sum_{i=1}^{3} a_i[1-e^{-\rho_a L\kappa_o(\lambda_i/\lambda)^{-n}}] \tag{5-13b}$$

$a_1 = 0.28$	$a_2 = 0.44$	$a_3 = 0.28$
$f_e(\lambda_1 T) = 0.11$	$f_e(\lambda_2 T) = 0.50$	$f_e(\lambda_3 T) = 0.89$
$\lambda_1 T = 2260$	$\lambda_2 T = 4110$	$\lambda_3 T = 8900\ \mu mK$

The 3-point or 3-wavelength approximation is equivalent to three gray bands advocated by Hottel [17], the bands here having wavelength limits corresponding to $f(\lambda T) = 0, 0.28$, $0.28+0.44 = 0.72$, and 1.

In contrast to the case of a cloud with a distribution of particle sizes, a molecular gas absorbs and emits in arrays of lines forming bands. The result is that the absorption coefficient varies very rapidly with wavenumber or wavelength, and the spectral absorption coefficient is the sum of the contributions of all the lines

$$\kappa_a(\nu) = \sum_i \kappa_i(\nu) \tag{5-14}$$

where each line has its own shape, intensity and location. For example, the Lorentz collision-broadened line shape dictates that

$$\kappa_i(\nu) = \frac{S_i \gamma_i}{\pi[(\nu-\nu_i)^2+\gamma_i^2]} \tag{5-15}$$

The quantity S_i (not to be confused with scattering source intensity; unfortunately the same symbols are used in the more-or-less separate bodies of literature) is the integrated intensity

$$S_i = \int_0^\infty \kappa_i(\nu)\,d\nu \doteq \int_{-\infty}^{+\infty} \kappa_i\,d(\nu-\nu_i) \tag{5-16}$$

The quantity γ_i is called the line half width, because

$$\kappa_i(\nu-\nu_i=\gamma_i) = \frac{1}{2}\,\kappa_i(\nu-\nu_i=0) \tag{5-17}$$

Typical values of γ at room temperature are on the order of

10^{-1} cm^{-1}. The line spacings $\nu_i - \nu_{i-1}$ are on the order of 10^0 or 10^1 cm^{-1}. The result is that κ_a may vary by several orders of magnitude over a few wavenumbers.

Usually the engineer or applied scientist prefers to make calculations with a more smoothly varying function. Hence, the gas emissivity $1-e^{-\rho_a \kappa_a L}$ or transmissivity $e^{-\rho_a \kappa_a L}$ is usually smoothed by averaging over a few line spacings

$$\bar{\tau} = \frac{1}{\Delta\nu} \int_{\nu-\Delta\nu/2}^{\nu+\Delta\nu/2} e^{-\rho_a \kappa_a L} d\nu \tag{5-18}$$

Because of the smoothing, the result does *not* in general obey the Beer-Lambert law, i.e.,

$$\bar{\tau} \neq e^{-\bar{\kappa}_a \rho_a L}$$

Instead it is necessary to prescribe an array of lines and carry out the integration prescribed by Eq. (5-18). The result for $\bar{\tau}$ is called a narrow band model. For example the Goody narrow band model gives

$$\bar{\tau} = \exp\left\{ \frac{-(S/d)X}{\sqrt{1+[(S/d)X/(\pi\gamma/d)]}} \right\} \tag{5-19}$$

where X is the absorber-partial-density-path-length product $\rho_a L$ and (S/d) and $\eta = \pi\gamma/d$ are narrow band parameters, the mean line-intensity-to-spacing ratio and mean line-width-to-spacing ratio respectively.

There are thus seen to be two "spectral" quantities, (1) the true spectral transmissivity τ that does obey the Lambert Beer Law but which fluctuates so rapidly with wavenumber that it is useless in practical calculations and (2) the spectrally smoothed transmissivity $\bar{\tau}$. The smoothed spectral emissivity $1-\bar{\tau}$ of a gas thus varies with X in the peculiar way prescribed by Eq. (5-19). When X is very small, the emissivity grows linearly with X. When X is somewhat larger, it may grow as $\sqrt{X}$. Finally, when X is very large it grows at $1-e^{-C\sqrt{X}}$ (according to this particular narrow band model). The line width varies with collision frequency and hence pressure and composition; thus when emissivity varies with $\sqrt{X}$ there is a $\sqrt{P_e}$ dependency as well. At high pressure when $\pi\gamma/d$ is large, the Beer-Lambert law does apply with (S/d) playing the role of $\bar{\kappa}_a$. The lines are said to be overlapped.

A wide-band model prescribes how (S/d) and η vary with wavenumber. The model can either be rational, i.e., a quantum-mechanical-based model, or arbitrary. Among the former is the harmonic-oscillator-rigid-rotator model which prescribes, approximately

$$(S/d) = (S/d)_o(\nu-\nu_o)e^{-(\nu-\nu_o)^2/\omega^2} \tag{5-20}$$

$$\eta = \beta P_e \tag{5-21}$$

Among the latter are the Schack model

$$(S/d) = \begin{cases} 0 & |\nu-\nu_o| > \omega/2 \\ (S/d)_o[1-2|\frac{\nu-\nu_o}{\omega}|] & |\nu-\nu_o| \leq \omega/2 \end{cases} \qquad (5\text{-}22)$$

$$\eta = \infty$$

and the exponential-tailed band model

$$(S/d) = (S/d)_o \exp[-2|\nu-\nu_o|/\omega] \qquad (5\text{-}23)$$

$$\eta = \beta P_e \qquad (5\text{-}24)$$

Equations (5-20 and 21) or (5-23 and 24) give a reasonably accurate model for representing measurements of gas properties, because they make provision for both line and band structures. A wide band model is thus seen to depend upon three parameters, the band width parameter ω, the line width parameter β, and the "integrated band intensity" α (not to be confused with absorptivity)

$$\alpha = \int_0^{\infty} (S/d)\,d\nu \doteq \int_{-\infty}^{+\infty} (S/d)\,d(\nu-\nu_o) \qquad (5\text{-}25)$$

which sets the value of $(S/d)_o$. The band origin ν_o locates the band spectrally.

For CO_2 bands ω_k for the kth absorption band is small. Even for H_2O bands it is not unreasonable to make the band approximation based upon the idea that the band origins are widely separated compared to the band widths, i.e., $\omega_k << \Delta\nu_k$. The band approximation is thus

$$\int_0^{\infty} [1-e^{-\rho_a \kappa_a L}] I_b(\nu) d\nu \doteq \sum_k I_b(\nu_k) \int_{-\infty}^{+\infty} [1-e^{-\rho_a \kappa_a L}] d(\nu-\nu_k) \tag{5-26}$$

The quantity with units of wavenumber is the band absorption

$$A_k = \int_{-\infty}^{+\infty} [1-e^{-\rho_a \kappa_a L}] d(\nu-\nu_k) \tag{5-27}$$

$$A_k = \int_{-\infty}^{+\infty} [1-\bar{\tau}] d(\nu-\nu_k) \tag{5-28}$$

When one introduces a wide band model one finds

$$A_k^* = A_k/\omega = F(\tau_{H,k}, \eta_k) \tag{5-29}$$

where $\tau_{H,k}$ is the "optical thickness at the band head" (for an exponential-tailed band model)

$$\tau_{H,k} = \alpha_k X/\omega_k \tag{5-30}$$

Section 5.E presents formulas, tabulated physical values, and graphs for finding $\tau_{H,k}$ and η_k for combustion gases H_2O, CO_2, CO, NO, SO_2 and the fuel CH_4. The values there are based upon collision-broadened lines. At very high temperatures and low pressure Doppler braodening would have to be considered for gases (such as HF and HCl) with small values of line width to spacing ratio.

The band absorption property is seen to be a spectrally integrated property that is nevertheless not a total property, because it is not integrated over the total spectrum. Total emissivity and absorptivity are defined on the basis of a total-spectrum integration. Consider a gas at temperature

T_g exposed to black body environment at temperature T_e. The total emissivity is

$$\varepsilon_g = \frac{\int_0^\infty [1-\bar{\tau}] I_b(\nu,T_g)\,d\nu}{\int_0^\infty I_b(\nu,T_g)\,d\nu} \tag{5-31}$$

$$\varepsilon_g \doteq \frac{\sum_k A_k I_b(\nu_k,T_g)}{\sigma T_g^{\,4}/\pi} = \sum_k b_k A_k = \sum_k b_k \omega_k A_k^* \tag{5-32}$$

where

$$b_k = \frac{\pi I_b(\nu_k,T_g)}{\sigma T_g^{\,4}} = \frac{B(\nu_k,T_g)}{\sigma T_g^{\,4}} \tag{5-33}$$

Similarly the total absorptivity is

$$\alpha_g = \frac{\int_0^\infty [I-\bar{\tau}] I_b(\nu,T_e)\,d\nu}{\int_0^\infty I_b(\nu,T_e)\,d\nu} \tag{5-34}$$

$$\alpha_g \doteq \sum_k b_k(T_e)\,\omega_k(T_g)\,A^*_{k,g} \tag{5-35}$$

Having a graph of calculated or experimentally measured total emissivity ε_g one often wishes to make a quick estimate of total absorptivity. Approximate rules for finding α_g from ε_g follow from Eqs. (5-32), (5-33), and (5-35). The recipe is to find a scaled length L', look up the emissivity at the environmental temperature, and correct it back to the gas temperature

$$\alpha_g(T_e,T_g,PL,P_e) \doteq (T_g/T_e)^{1/2}\varepsilon_g(T_e,PL',P_e) \qquad (5\text{-}36)$$

The rationale of finding ε_g at temperature T_e is that the values of $b_k(T_e)$ are correct. The correction of $(T_g/T_e)^{1/2}$ is based upon the fact that $\omega_k(T_g)$ varies as $T_g^{1/2}$. The recipe for L' must account for the variation of both $\tau_{H,k}$ and η_k with temperature. At high temperature and pressure the lines are overlapped so that the large values of η_k are no consequence. Thus one wants

$$\tau_{H,k} = \alpha_k(T_g)\rho_a(T_g)L/\omega_k(T_g) = \tau'_{H,k}$$
$$= \alpha_k(T_e)\rho_a(T_e)L'/\omega_k(T_e) \qquad (5\text{-}37)$$

For the important fundamental bands α_k is constant in T_g; thus we obtain Penner's Rule

$$L' = L\cdot(T_e/T_g)^{3/2} \qquad (\eta_k > 1) \qquad (5\text{-}38)$$

At moderate temperatures and atmospheric pressure A_k^* depends upon $\eta_k\tau_{H,k}$, and η_k goes approximately as $T_g^{1/2}$ ($300 \leq T_g \leq 1200$ K). Thus we obtain Hottel's Rule

$$L' = L\cdot(T_e/T_g) \qquad (\eta_k < 1,\ \tau_{H,k} > 1/\eta) \qquad (5\text{-}39)$$

Figures 5.6 and 5.7 present total emissivity charts for gaseous CO_2 and H_2O respectively. The charts pertain to a dilute mixture of the radiating species in a mixture with nitrogen at a total pressure of one atmosphere. Section 5.E explains how the charts were calculated and how to treat other values of total pressure. In the range where Eq. (5-38)

FIGURE 5-6

Total Emissivity of CO_2 Gas (from [19])

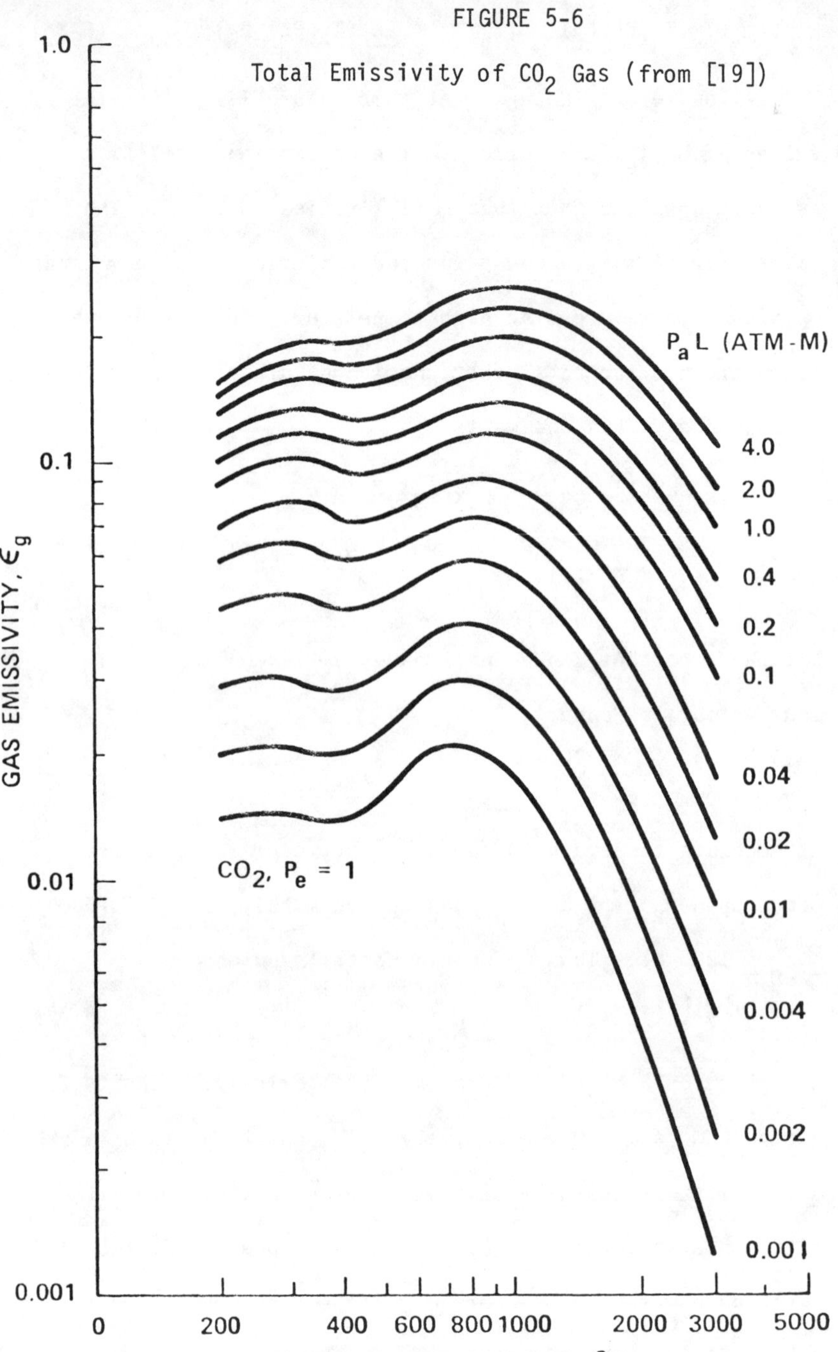

FIGURE 5-7

Total Emissivity of H_2O Gas (from [19])

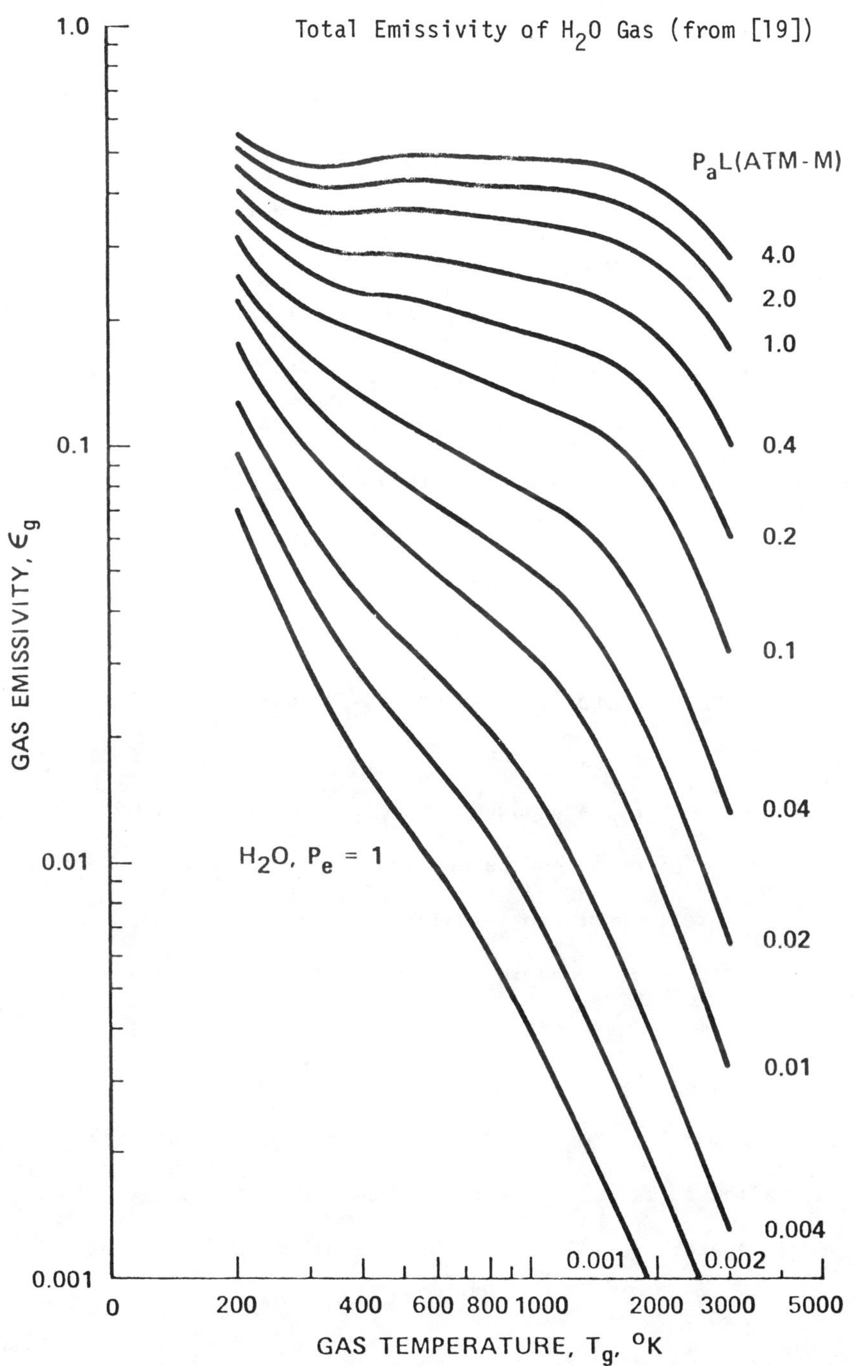

applies, there is no pressure correction. In the range where Eq. (5-39) applies $L\cdot(T_e/T_g)$ is multiplied by P_e, i.e.,

$$L' = L\cdot(T_e/T_g)\cdot P_e \tag{5-40}$$

This equation and Eq. (5-39) with $T_e = T_g$ apply to the emissivity as well. The dimensionless equivalent broadening pressure P_e is found from Eq. (5-57).

Section 5.F deals briefly with gas mixtures.

5.E Molecular Gas Radiation Properties. Given a narrow band model such as Eq. (5-19) and a wide band model such as Eqs. (5-23,24), one needs values of band width ω_k, line width to spacing parameter β_k, and band intensity α_k or optical thickness $\tau_{H,k} = \alpha_k X/\omega_k$. With these values and Eqs. (5-19) and (5-23,24) one can find spectral properties or one can find band absorption via Eqs. (5-28) and (5-32,35). The determination of ω_k, β_k, and α_k or $\tau_{H,k}$ and the specification of F in Eq. (5-28) are now described.

Table 5-1 contains the necessary information. Shown for each gas are the principal radiating bands such as the 6.3 μm H_2O band. Each band is described by a set of vibrational quantum number increments $\pm\delta_1, \pm\delta_2, \pm\delta_3, \ldots$, such as 0,1,0 for the 6.3 μm H_2O band. The number of vibrational degrees of freedom, m = 3 for H_2O, and the wavenumber ν_i and statistical weight g_i associated with each is shown. A band with a single nonzero positive value of δ_i equal to unity is called

TABLE 5-1

EXPONENTIAL WIDE BAND MODEL PARAMETERS[a]

Gas	Vibrations ν_h (cm^{-1})	Band $\delta_1,\delta_2,\ldots$	Pressure Parameters n	Pressure Parameters b (T_o = 100 K)	Spectral Location ν_i (cm^{-1})	Spectral Location ν_c (cm^{-1})	Spectral Location ν_n (cm^{-1})	Band Absorption Parameters α_o ($\frac{cm^{-1}}{gm\ m^{-2}}$)	Band Absorption Parameters β_o	Band Absorption Parameters ω_o (cm^{-1})
(1) H_2O	m = 3	(1) Rotational								
	ν_1 = 3652		1	$8.6(T_o/T)^{½}+0.5$	0[b]			5200.0[b]	0.14311[b]	28.4[b]
	ν_2 = 1595	0,0,0								
	ν_3 = 3756	(2) 6.3 µm		$8.6(T_o/T)^{½}+0.5$		1600		41.2	0.09427	56.4
	g_1 = 1	0,1,0	1							
	g_2 = 1	(3) 2.7 µm								
	g_3 = 1	0,2,0						0.19		
		1,0,0	1	$8.6(T_o/T)^{½}+0.5$		3760		2.30	0.13219	60.0
		0,0,1						22.40		
		(4) 1.87 µm	1	$8.6(T_o/T)^{½}+0.5$		5350		3.0	0.08169	43.1
		0,1,1								
		(5) 1.38 µm	1	$8.6(T_o/T)^{½}+0.5$		7250		2.5	0.11628	32.0
		1,0,1								
(2) CO_2[d]	m = 3	(1) 15 µm								
	ν_1 = 1351	0,1.0	0.7	1.3		667		19.0	0.06157	12.7
	ν_2 = 667	(2) 10.4 µm								
	ν_3 = 2396	-1,0,1	0.8	1.3		960		$2.47x10^{-8}$	0.04017	13.4
	g_1 = 1	(3) 9.4 µm								
	g_2 = 2	0,-2,1[c]	0.8	1.3		1060		$2.48x10^{-9}$[c]	0.11888[c]	10.1
	g_3 = 1	(4) 4.3 µm								
		0,0,1	0.8	1.3			2410	110.0	0.24723	11.2
		(5) 2.7 µm								
		1,0,1	0.65	1.3		3660		4.0	0.13341	23.5
		(6) 2.0 µm								
		2,0,1	0.65	1.3		5200		0.066	0.39305	34.5
(3) CO	m = 1	(1) 4.7 m								
	ν_1 = 2143	1	0.8	1.1		2143		20.9	0.07506	25.5
	g_1 = 1	(2) 2.35 m								
		2	0.8	1.0		4260		0.14	0.16758	20.0

Table 5-1 (continued)

Gas	Vibrations ν_{h_1} (cm^{-1})	$\delta_1, \delta_2, \ldots$	Pressure Parameters n	b (T_o = 100 K)	Spectral Location ν_i (cm^{-1})	ν_c (cm^{-1})	ν_n (cm^{-1})	Band Absorption Parameters α_o ($\frac{cm^{-1}}{gm\ m^{-2}}$)	β_o	ω_o (cm^{-1})
(4) NO	m = 1	(1) 5.34 μm								
	ν_1 = 1876	1	0.65	1.0		1876		9.0	0.18050	20.0
	g_1 =									
(5) SO_2	m = 3	(1) 19.27 μm								
	ν_1 = 1151	0,1,0	0.7	1.28		519		4.22	0.05291	33.08
	ν_2 = 519	(2) 8.68 μm								
	ν_3 = 1361	1,0,0	0.7	1.28		1151		3.674	0.05952	24.83
	g_1 = 1	(3) 7.35 μm								
	g_2 = 1	0,0,1	0.65	1.28		1361		29.97	0.49299	8.78
	g_3 = 1	(4) 4.34 μm								
		2,0,0	0.6	1.28		2350		0.423	0.47513	16.45
		(5) 4.0 μm								
		1,0,1	0.6	1.28		2512		0.346	0.58937	10.91
(6) CH_4	m = 4	(1) 7.66 μm								
	ν_1 = 2914	0,0,0,1	0.8	1.3		1310		28.0	0.08698	21.0
	ν_2 = 1526	(2) 3.31 μm								
	ν_3 = 3020	0,0,1,0	0.8	1.3		3020		46.0	0.06973	56.0
	ν_4 = 1306	(3) 2.37 μm								
	g_1 = 1	1,0,0,1	0.8	1.3		4220		2.9	0.35429	60.0
	g_2 = 2	(4) 1.71 μm								
	g_3 = 3	1,1,0,1	0.8	1.3		5861		0.42	0.68598	45.0
	g_4 = 3									

[a] From Edwards and Balakrishnan [15].

[b] For the rotational bands of H_2O $\alpha(T) = \alpha_o$ and $\beta(T) = \beta_o(T/T_o)^{-\frac{1}{2}}$. Otherwise α_o, β_o and ω_o apply to equations (43), (41), and (47).

[c] Because of Fermi resonance between the ν_1 and $2\nu_2$ levels, the Ψ and Φ functions for the 1060 cm^{-1} band are to be those of the 960 cm^{-1} band; i.e. use the set of δ's for the 960 cm^{-1} band to get Ψ and Φ for either band.

[d] The 1,0,0 band of the linear CO_2 molecule appears only at high pressures when a dipole is induced by collisions.

a fundamental band, and unless a symmetric vibration is associated with it, is strong. The symmetric band, such as the CO_2 1,0,0 has negligible intensity except at high pressure, where it is said to have a pressure-induced (collision-induced) dipole. Thus the symmetric diatomic gases O_2, N_2, H_2, etc. do not appear in the table. A band with a single nonzero value of δ_i equal to 2 or more is called an overtone and is weak, often enough so to be neglected. A band with two or more nonzero values of δ_i is called a combination band, such as the 9.4 μm CO_2 band.

The band spectral location is in the vicinity of the sum of the $\pm\delta_i\nu_i$ products. The spectral location recommended for calculation is shown in the table as either a band center ν_c for an approximately symmetric band or as an upper limit ν_u for the 4.3 μm band which has a "head" at approximately 2410 cm^{-1}. Numerical values are shown for pressure parameters n and b and band absorption parameters α_o, β_o, and ω_o.

The band width parameter ω_k for the kth absorption band is readily calculated from the values given in the table. It is simply

$$\omega_k = \omega_o(T/T_o)^{1/2} \tag{5-41}$$

where ω_o is related to the rotational constant B (half of the line spacing d) by

$$\omega_o = 0.9\ \Gamma^2(\tfrac{3}{4})(2kT_o/hcB)^{1/2} \tag{5-42}$$

where T_o = 100 K has been chosen for the table. Thus for the 6.3 μm H_2O band at 1000 K, $\omega = 56.4(1000/100)^{1/2}$ = 178.4 cm^{-1}. The quantity α_k is the value α_o for a fundamental band, but in general for a combination or overtone band .

$$\alpha_k(T) = \alpha_o \frac{[1-\exp(-\sum_{i=1}^{m} \pm u_i \delta_i)]\Psi(T)}{[1-\exp(-\sum_{i=1}^{m} \pm u_{o,i} \delta_i)]\Psi(T_o)} \tag{5-43}$$

where

$$u_i = hc\nu_i/kT \quad , \quad hc/k = 14388 \ \mu m \ K \tag{5-44}$$

and

$$\Psi(T) = \frac{\prod_{i=1}^{m} \sum_{v_i - v_{o,i}}^{\infty} \frac{(v_i+g_i+\delta_i-1)!}{(g_i-1)!v_i!} e^{-u_i v_i}}{\prod_{i-1}^{m} \sum_{v_i=0}^{\infty} \frac{(v_i+g_i-1)!}{(g_i-1)!v_i!} e^{-u_i v_i}} \tag{5-45}$$

$$v_{o,i} = \begin{cases} 0 & \text{when a positive sign is associated with } \delta_i \text{, or } \delta_i=0. \\ \delta_i & \text{when a negative sign is associated with } \delta_i. \end{cases} \tag{5-46}$$

Note that the values of δ_i shown in the equations are understood to be the absolute values unless preceded by ±; then the appropriate sign is to be associated with δ_i.

The line-width-to-spacing parameter β is given by

$$\beta(T) = \beta_o(T/T_o)^{-1/2}\Phi(T)/\Phi(T_o) \tag{5-47}$$

where

$$\Phi(T) = \frac{\left\{ \prod_{i=1}^{m} \sum_{v_i=v_{o,i}}^{\infty} \left[\frac{(v_i+g_i+\delta_i-1)!}{(g_i-1)!v_i!} e^{-u_i v_i} \right]^{1/2} \right\}^2}{\prod_{i=1}^{m} \sum_{v_i=v_{o,i}}^{\infty} \frac{(v_i+g_i+\delta_i-1)!}{(g_i-1)!v_i'} e^{-u_i v_i}} \tag{5-48}$$

Notice that the 2.7 μm H_2O band is composed of three superposed bands, in order of importance, the (0,0,1), (1,0,0), and (0,2,0) bands. For such a band it is necessary to compute three values of α_j and three values of β_j and then combine them

$$\alpha_k = \sum_{j=1}^{3} \alpha_{k,j} \tag{5-49}$$

$$\beta_k = \frac{1}{\alpha_k} \left\{ \sum_{j=1}^{3} (\alpha_{k,j}\beta_{k,j})^{1/2} \right\}^2 \tag{5-50}$$

Because the formulas for computing Ψ and Φ are rather complex, convenient graphs giving $\tau_{H,k}$ and β_k are presented in Figs. 5-8 through 5-11.

The equivalent broadening pressure P_e is based upon a simplified kinetic theory of molecules with optical collision diameters, D_a for the absorbing species, and D_b for the braodening species taken here to be nitrogen. The line width γ results from the natural lines being shifted spectrally during molecular collisions and is proportional to the collision frequency $F_{aa} + F_{ab}$ where

$$F_{ab} = \Delta v \pi D_{ab}^2 (P_b/kT) \tag{5-51}$$

FIGURE 5-8

Optical Depth at Band Head, CO and CO_2 from [15]

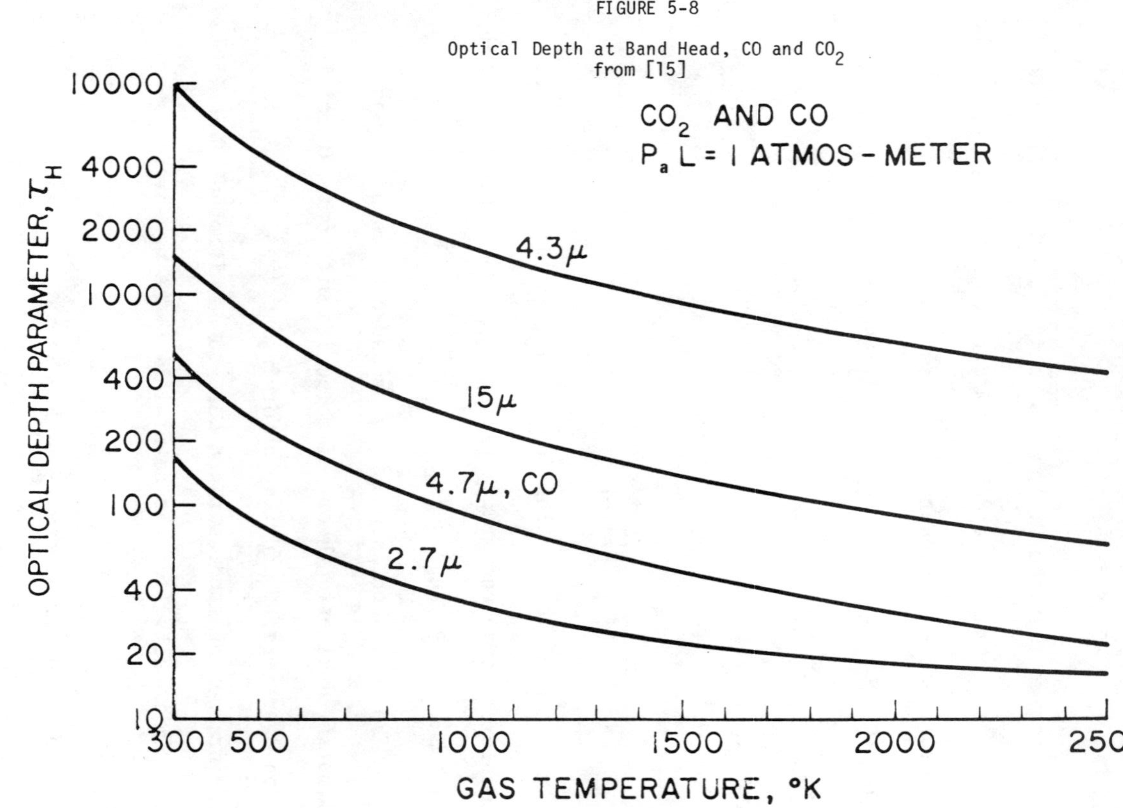

FIGURE 5-9

Line Width Parameter, CO and CO_2 from [15]

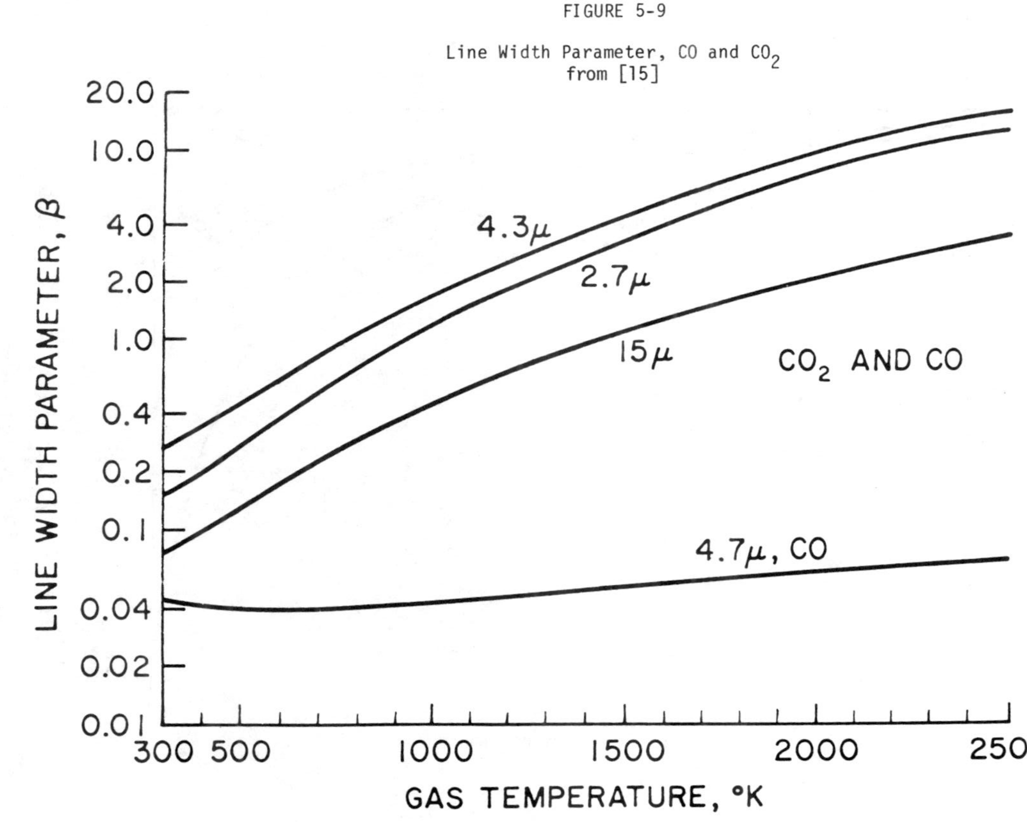

FIGURE 5-10

Optical Depth at Band Head, H_2O
from [15]

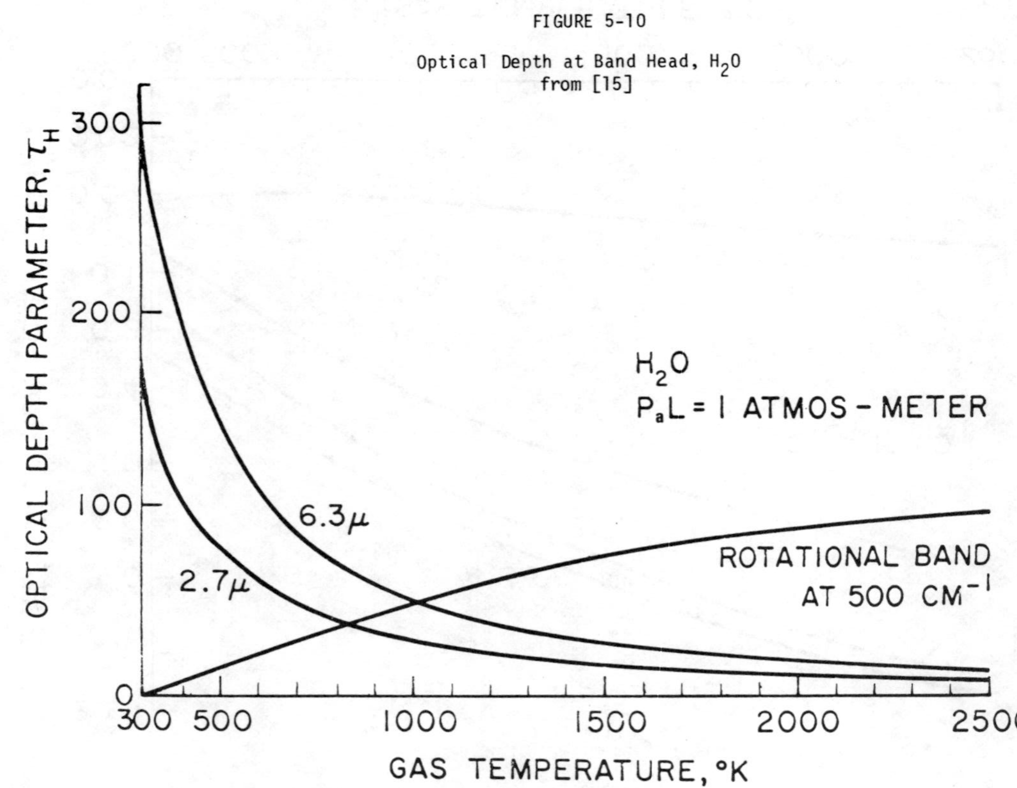

FIGURE 5-11

Line Width Parameter, H_2O
from [15]

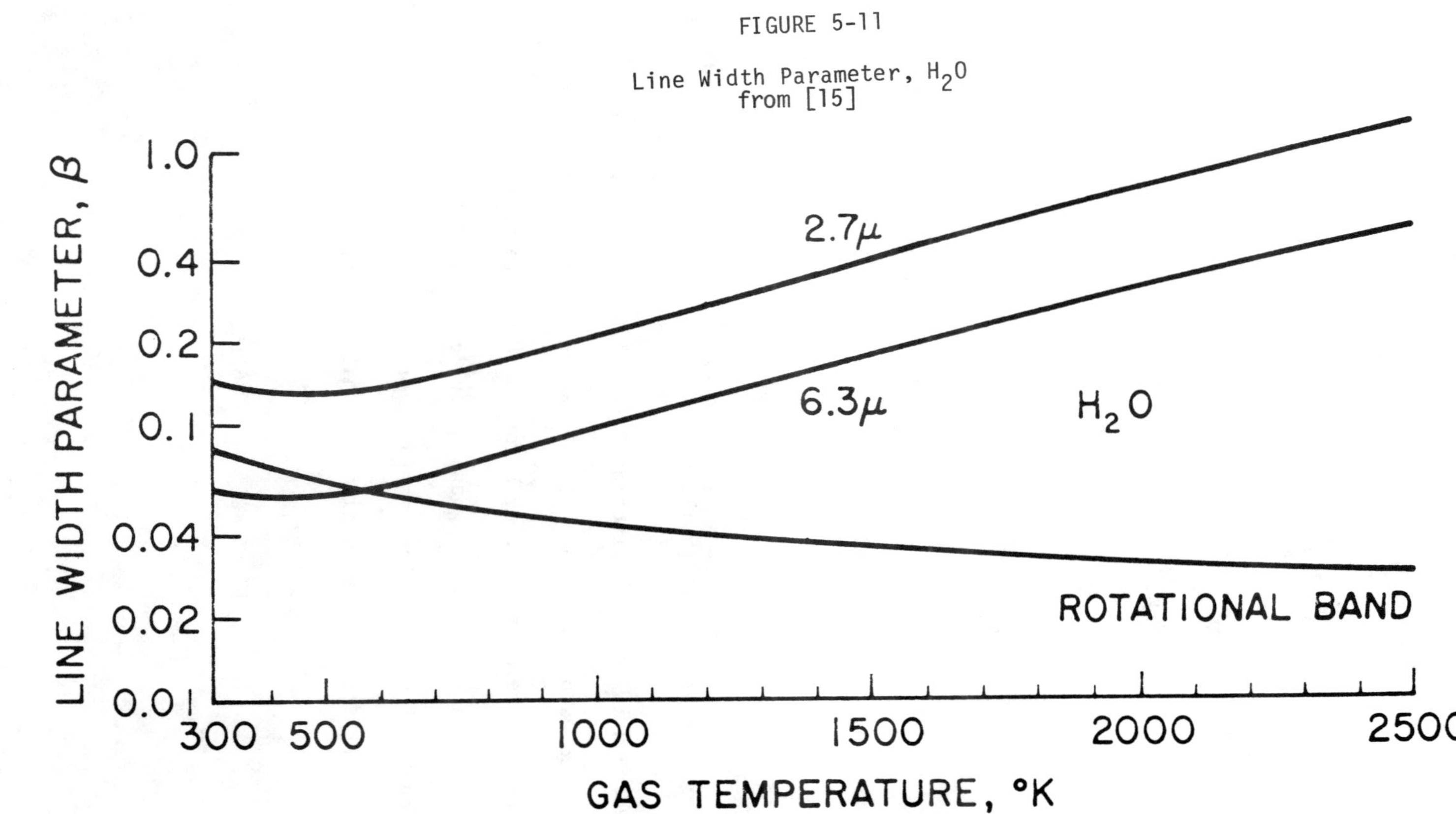

where

$$D_{ab} = \frac{1}{2} [D_a + D_b] \tag{5-52}$$

$$\Delta v = \left(\frac{16kT}{\pi}\right)^{1/2} \left[\frac{1}{m_a} + \frac{1}{m_b}\right]^{1/2} \tag{5-53}$$

Thus one expects that

$$\gamma = \gamma_o (T/T_o)^{-1/2} [(P_b + bP_a)/P_o] \tag{5-54}$$

where

$$b = \frac{D_{aa}^2}{D_{ab}^2} \left(\frac{2m_b}{m_a + m_b}\right)^{1/2} \tag{5-55}$$

On empirical grounds the expression is adjusted by exponent n. Thus η is written

$$\eta = \pi\gamma/d = \beta P_e \tag{5-56}$$

where

$$P_e = \{[P/P_o][1 + (b-1)x]\}^n \tag{5-57}$$

where x is the mole fraction P_a/P. Table 5.1 gives recommended values of n and b for $P_o = 1$ atm. Note the strong temperature dependence of b for the H_2O molecules.

With values of $\tau_{H,k}$ and η_k one can find A_k^* from the four-region expressions as follows:

(1) The linear region for small optical depth and high pressure,

$$\tau_{H,k} \leq 1 \qquad \tau_{H,k} \leq \eta_k \qquad A_k^* = \tau_{H,k} \tag{5-58a}$$

(2) The square root region for small to moderate optical depth and low pressure,

$$\eta_k \le \tau_{H,k} \le 1/\eta_k \ , \quad \eta_k \le 1 \ , \quad A_k^* = (4\eta_k\tau_{H,k})^{1/2} - \eta_k \tag{5-58b}$$

(3) The log-root region for large optical depth and low pressure,

$$1/\eta_k < \tau_{H,k} \ , \quad \eta_k \le 1 \ , \quad A_k^* = \ln(\tau_{H,k}\eta_k) + 2 - \eta_k \tag{5-58c}$$

(4) The log region for large optical depth and high pressure,

$$\tau_{H,k} \ge 1 \ , \quad \eta_k \ge 1 \ , \quad A_k^* = \ln\tau_{H,k} + 1 \tag{5-58d}$$

When the approximate 4-region expressions are compared with exact numerical integration of Eq. (5-28), they are found to be higher. However, the values in Table 5-1 were found by correlating the approximate 4-region expressions to band absorption measurements; thus use of the 4-region expressions gives results that agree (±15 per cent) with physical reality. If one wants to employ Eq. (5-23) for approximate spectral calculations, it is recommended to increase the value of ω_k obtained from the table by 20 per cent.

Even when one intends to use a total emissivity chart, it is well to calculate $\tau_{H,k}$ and η_k of the two or three most important bands in order to know whether to apply Eq. (5-38) or (5-39) to obtain total absorptivity and to know whether or not to apply the pressure correction according to Eq. (5-40). Furthermore, one needs to know where in Table 6.1 to find the relation between beam length and geometric mean beam length as described in Section 6.2.

5.F Gas Mixtures. For overlapping bands of the *same species* the line structure is similar, and one employs Eqs. (5-49) and (5-50). For overlapping bands of different species, one expects that there is no correlation in the line structure so that the product of the individual transmissivities, each given by Eq. (5-18), applies on a spectral basis.

$$\bar{\tau}_{\nu,mix} = \Pi\, \bar{\tau}_{\nu,i} \qquad \text{(i is the species)} \qquad (5\text{-}59)$$

$$\bar{\varepsilon}_{\nu,mix} = \bar{\alpha}_{\nu,mix} = 1-\tau_{\nu,mix} \qquad (5\text{-}60)$$

If one wants to make a hand calculation and avoid detailed spectral calculations, one uses a block calculation procedure as follows: (1) Determine for each band of each absorbing-emitting species the band absorption A_i. (2) Assign to each band a band transmissivity $\tau_{g,i}$ based upon

$$\tau_{g,i} = (\tau_{H,i}/A_i^*)(dA_i^*/d\tau_{H,i}) \leq 0.90 \qquad (5\text{-}61)$$

where A_i^* comes from Eqs. (5-58a,b,c,or d). If the indicated transmissivity from Eq. (5-61) is greater than 0.90, the value

0.90 is used. (3) Find the band width

$$\Delta\nu_i = A_i/(1-\tau_{g,i}) \tag{5-62}$$

(4) Following recommended center or upper-limit locations, locate the bands. (5) Order the band limits in ascending order of wavenumber, starting with zero, and find the transmissivity in each block between limits. For the kth block the transmissivity is

$$\tau_{g,k} = \Pi\ \tau_{g,i} \tag{5-63}$$

where the product is over all bands within the block (the bands outside the block have a transmissivity of 1.0 within the block). (6) Use the blocked out $\tau_{g,k}$ as desired to calculate total transfer factors or total absorptivity and emissivity. For example for total emissivity

$$\varepsilon_g = \int_0^1 [1-\tau_g]df_e(T_g/\nu) \tag{5-64}$$

$$\varepsilon_g = \sum_{k=1}^{n} [1-\tau_{g,k}][f_e(T_g/\nu_k) - f_e(T_g/\nu_{k+1})] \tag{5-64}$$

Note that the block procedure outlined here is preferable to using Eq. (5-32) or (5-35), even when a pure species (with or without transparent broadener diluent gas) is involved, when the absorption bands are wide.

Tables 5-2 and 5-3 from Ref. [19] show a sample block calculation.

TABLE 5-2

BAND ABSORPTIONS FOR A GAS MIXTURE
(0.18 H_2O, 0.06 CO_2, 0.03 CO, 0.73 N_2, P=1 atm, L=3 m, T_g=1400°K)
from [19]

Species i	Gas	j	Band Region	Band Absorption $A_{i,j}$ cm^{-1}	Band Transmissivity $\tau_{g,i,j}$	Band Width $\Delta\nu_{i,j}$ cm^{-1}	Lower Limit $\nu_{\ell,i,j}$ cm^{-1}	Upper Limit $\nu_{u,i,j}$ cm^{-1}
1	H_2O	1	Rotational	775	0.137	898	0	898
		2	6.3 μ	638	0.331	954	1123	2077
		3	2.7 μ	676	0.332	1012	3254	4266
		4	1.87 μ	171	0.594	422	5139	5561
		5	1.38 μ	131	0.610	336	7082	7418
2	CO_2	1	15 μ	205	0.232	266	534	800
		2	10.4 μ	10	0.90	100	910	1010
		3	9.4 μ	10	0.90	100	1010	1060
		4	4.3 μ	260	0.161	310	2100	2410
		5	2.7 μ	220	0.400	366	3477	3843
		6	1.9 μ	9	0.90	90	5155	5245
3	CO	1	4.7 μ	89	0.527	188	2049	2237
		2	2.35 μ	3	0.90	34	4243	4277

TABLE 5-3

BLOCK CALCULATION OF TOTAL EMISSIVITY AND ABSORPTIVITY
(0.18 H_2O, 0.06 CO_2, 0.03 CO, 0.73 N_2, P=1 atm, L=3 m, T_g=1400°K, T_e=1100°K)
from [19]

Block Number k	Lower Limit ν_k	Upper Limit ν_{k+1}	Block Transmissivity $\tau_{g,k}$	Fractional Function $f(T_g/\nu_k)$	Block Fraction $\varepsilon_{g,k}\Delta f_k$	Fractional Function $f(T_s/\nu_k)$	Block Fraction $\alpha_{g,k}\Delta f_k$
1	0	534	0.137	1.000	0.006	1.000	0.013
2	534	800	0.032	0.993	0.014	0.985	0.023
3	800	898	0.137	0.979	0.007	0.961	0.011
4	898	910	1.000	0.971	0.000	0.948	0.000
5	910	1060	0.900	0.970	0.001	0.946	0.003
6	1060	1123	1.000	0.957	0.000	0.920	0.000
7	1123	2049	0.331	0.950	0.102	0.909	0.156
8	2049	2077	0.174	0.797	0.004	0.676	0.007
9	2077	2100	0.527	0.792	0.002	0.668	0.003
10	2100	2237	0.085	0.787	0.027	0.662	0.036
11	2237	2410	0.161	0.758	0.033	0.623	0.040
12	2410	3254	1.000	0.719	0.000	0.575	0.000
13	3254	3477	0.332	0.534	0.032	0.357	0.032
14	3477	3843	0.133	0.486	0.063	0.309	0.058
15	3843	4243	0.332	0.413	0.049	0.242	0.040
16	4243	4266	0.299	0.339	0.003	0.182	0.002
17	4266	4277	0.335	0.335	0.002	0.179	0.001
18	4277	5139	1.000	0.332	0.000	0.177	0.000
19	5139	5155	0.594	0.211	0.001	0.090	0.000
20	5155	5245	0.535	0.209	0.005	0.089	0.003
21	5245	5561	0.594	0.199	0.014	0.083	0.008
22	5561	7082	1.000	0.165	0.000	0.064	0.000
23	7082	7418	0.610	0.063	0.005	0.016	0.002
24	7418	∞	1.000	0.050	0.000	0.012	0.000
					0.370		0.438

REFERENCES TO SECTION 5

I. The Equation of Transfer

1. Davidson, N., Statistical Thermodynamics, McGraw-Hill, New York, 1962, pp. 220-226.

2. Goody, R. M., Atmospheric Radiation, Oxford, London, 1964, pp. 17-44.

3. Zeldovich, Y. B., and Y. P. Razier, Physics of Shock Waves and High-Temperature Hydrodynamic Phenomena, Academic Press, New York, 1966, Vol. I, pp. 107-133.

II. Reviews of Experimental Measurement Techniques

4. Hottel, H. C., and A. F. Sarofim, Radiation Transfer, McGraw-Hill, New York, 1967, Chapter 6.

5. Edwards, D. K., "Thermal Radiation Measurements", Ch. 9 of Measurements in Heat Transfer, E. R. G. Eckert and R. J. Goldstein editors, Hemisphere/McGraw-Hill, New York, 1976. See also Ref. [19].

III. Properties of Particulates (see also Ref. 3, Chapter 7)

6. Van de Hulst, H. C., Light Scattering by Small Particles, Wiley, New York, 1957.

7. Hottel, H. C., and A. F. Sarofim, Radiative Transfer, McGraw-Hill, New York, 1967, Chapter 12. See also Sarofim, A. F., and H. C. Hottel, "Radiative Transfer in Combustion Chambers: Influence of Alternative Fuels", Sixth International Heat Transfer Conference, Toronto, Canada, Aug. 7-11, 1978, Vol. 6, pp. 199-217.

8. Howarth, C. R., P. J. Foster, and M. W. Thring, "The Effect of Temperature on the Extinction of Radiation by Soot Particles", Proceedings of the Third International Heat Transfer Conference, Chicago, Aug. 7-12, 1966, Vol. 5, pp. 122-128.

9. Penner, S. S., and D. B. Olfe, Radiation and Reentry, Academic Press, New York, 1968, Chapter 4.

10. Hubbard, G. L., and C. L. Tien, "Infrared Mean Absorption Coefficients of Luminous Flames and Smoke", J. Heat Transfer, Vol. 100, pp. 235-239, 1978.

IV. Gas Properties

A. Narrow Band Models

11. Plass, G. N., "Useful Representation for Measurements of Spectral Band Absorption", J. Optical Soc. Am., Vol. 50, pp. 868-875, 1960.

12. Tiwari, S. N., and S. K. Gupta, "Accurate Spectral Modeling for Infrared Radiation", J. Heat Transfer, Vol. 100, pp. 240-246, 1978.

B. Wide Band Models

13. Schack, A., Industrial Heat Transfer, Wiley, New York, 1927.

14. Edwards, D. K., and W. A. Menard, "Comparison of Models for Correlation of Total Band Absorption", Applied Optics, Vol. 3, pp. 621-625, 1964.

15. Edwards, D. K., and A. Balakrishnan, "Thermal Radiation by Combustion Gases", International J. Heat and Mass Transfer, Vol. 16, pp. 25-40, 1973.

16. Chu, H. K., and R. Grief, "Theoretical Determination of Band Absorption for Nonrigid Rotation with Applications to CO, NO, N_2O, and CO_2", J. Heat Transfer, Vol. 100, pp. 230-234, 1978.

C. Total Emissivity Charts

17. Hottel, H. C., "Radiant Heat Transmission", Chapter 4 of W. H. McAdams' Heat Transmissions, McGraw-Hill, New York, 3rd Edition, 1954.

18. Eckert, E. R. G., and R. M. Drake, Analysis of Heat and Mass Transfer, McGraw-Hill, New York, 1972, pp. 653-656.

19. Edwards, D. K., "Molecular Gas Band Radiation", Advances in Heat Transfer, Vol. 12, pp. 115-193, 1976.

D. Ultra High Temperature Gases (see also Ref. 9, Chapters 1 and 3)

20. Menard, W. A., G. M. Thomas, and T. M. Helliwell, "Experimental and Theoretical Study of Molecular, Continuum, and Line Radiation from Planetary Atmospheres", AIAA Journal, Vol. 6, pp. 655-664, 1968.

EXERCISES FOR SECTION 5

1. A 10 m^3 first-stage combustion chamber is fed 1 kg/s of saw dust waste. The dust particles may be idealized as opaque spheres 1 mm in diameter at a pyrolyzing temperature of 200°C, with a solid density of 800 kg/m^3, and with a residence time in the chamber of 20 ms. Between the dust particles is a sooty flame at 1200°C with an absorption coefficient of 1 m^{-1}. (a) How many dust particles are present per unit volume? What absorption coefficient do they cause? (b) Write the equation of transfer for the soot and dust mixture. Solve it for the case of uniform composition. (c) What is the equivalent gray body temperature of the mixture and total absorption coefficient?

2. Shock-heated gas at 10^4 K lies between a 20 cm diameter hemisphere at 2000 K and a concentric transparent shock wave with a 15 cm radius of curvature. The hot gas has a total absorption coefficient of 1×10^{-3} cm^{-1}. (a) What is the total intensity of the irradiation on the stagnation point normal to the surface? (b) What is the total intensity of the irradiation at grazing incidence? (c) What is the irradiation on the stagnation point?

3. A two-meter diameter black sphere contains a 2000 K gas with absorption coefficient $k = 1\ m^{-1}$. The upper hemisphere wall is 1500 K, and the lower hemisphere is 1200 K. What is the net radiant flux at the lowest point of the bottom hemisphere?

4. A gas has a spectral absorption coefficient that varies according to $k = k_o(\nu/\nu_o)^{-3}$ when $\nu > \nu_o$ and is zero for $\nu < \nu_o$, where $hc\nu_o = 13.5$ eV. For a gas temperature of 10^5 K and a certain beam length L, the value of k_oL is 0.0015. Find the total emissivity of the gas beam.

5. A gas has three major bands, one centered at 667 cm^{-1}, one with an upper limit at 2410 cm^{-1}, and the third centered at 3660 cm^{-1}. For a beam length of 3 m, a partial pressure of 0.03 atm, a total pressure of 1 atm, and a temperature of 1400 K, the band absorptions are found to be $A_1 = 205\ cm^{-1}$, $A_2 = 260\ cm^{-1}$, and $A_3 = 220\ cm^{-1}$. (a) What is the total emissivity of the gas? (b) What is the total absorptivity for black body radiation from a 1100 K source?

6. Use the information in Table 5-1 for the 4.3 μm band of CO_2 and for the conditions shown on Figure 5-4 of the *Notes* ($T_g = 1389$ K, $P_{Total} = P_{CO_2} = 10$ atm, and $L = 0.387$ m) to do the following:

(a) Find $\psi(T_g)$ and the integrated intensity $\alpha(T_g)$

6. (continued)

   (b) Find $\omega(T_g)$ and τ_H. Compare your value of τ_H with a value from Fig. 5-8.

   (c) Find $\Phi(T_g)$ and the line width parameter $\beta(T_g)$. Compare with Fig. 5-9.

   (d) Find P_e and η

   (e) Find A* and A. Compare A with the area under the curve in Fig. 5-4.

7. Repeat Exercise 6 for the 2.7 μm band of H_2O for the conditions T_g = 833 K, $P_{H_2O} = P_{Total}$ = 2 atm, and L = 0.387 m. Note that three bands overlap, and Eqs. (5-49) and (5-50) apply. Discuss briefly the differences between the 2.7 μm H_2O band and the 4.3 μm CO_2 band.

8. Calculate and plot on a linear graph the mass absorption coefficient (in m^2/kg) for $2100 \leq \nu \leq 2110$ cm^{-1} assuming a harmonic-oscillator rigid-rotator representation of carbon monoxide gas as follows:

$$(S/d)_i \doteq \alpha[(\nu-\nu_i)/\omega^2]\exp[-(\nu-\nu_i)^2/\omega^2]$$

$$d = 2B,\ \omega = [4BkT/hc]^{\frac{1}{2}},\ \gamma = \gamma_o(P_e/P_o)(T/T_o)^{-\frac{1}{2}}$$

$$\nu_i = \nu_o - 2Bi\ ;\quad i = 1,2,3,\ldots$$

$$\kappa = \sum_{i=-\infty}^{+\infty} \frac{S_i\gamma_i/\pi}{(\nu-\nu_i)^2+\gamma_i^2}$$

Take T = 300 K, P_e/P_o = 1.3, α_o = 20.9 $cm^{-1}/(g\ m^{-2})$,

8. (continued)

$B = 1.931$ cm^{-1}, $\gamma_o = 0.08$ cm^{-1}, $\nu_o = 2143$ cm^{-1}. Plot the spectral transmissivity for a column of the gas with $X = \rho_a L = 1$, 10, and 100 g/m^2. Make a logarithmic cross-plot of the mean absorptivity versus X. Find how the curve for $P_e/P_o = 2.6$ plots on the cross-plot.

9. Hydrogen chloride gas has a fundamental vibration-rotation band centered at $\nu_1 = 2886$ cm^{-1}. The integrated intensity, band-width parameter, and line-width parameters are given approximately by $\alpha_1 = 9.2$ cm^{-1}/(g m^{-2}), $\omega = 89(T/T_o)^{\frac{1}{2}}$, $T_o = 300$ K, $\beta = 0.0188(T/T_o)^{-\frac{1}{2}}[1+0.0487(T/T_o)^{3/2}]^2$

   (a) Calculate the total emissivity of a 40 per cent (mol) mixture of HCl in a diluent at a total pressure of 2 atm for a beam length of 2.9 m. (Take n = b = 1 in Eq. (5-57) for lack of better information).

   (b) Calculate a curve of total emissivity versus gas temperature for the range of T_g on Fig. 5-6 and $P_a L = 1$ atm-m. Discuss the differences in shape between the calculated curve and that for CO_2 gas in the figure.

   (c) What "scaling rule" should be used to find total absorptivity from a total emissivity chart for HCl?

   (d) What rule should be used to find emissivity at a lower pressure from a one-atmosphere chart?

10. Use Figs. 5-8 and 5-9 to calculate the contribution of the 2.7, 4.3, and 15 μm bands of CO_2 to the total emissivity of 1 atm-m of the gas at $P_e = 1$ and $T = 1000$ K. Compare your result with the total emissivity in Fig. 5-6.

11. Calculate directly, from Eq. (5-35) (or Eq. (5-35) put in the form of Eq. (5-64)) and results of Exercise 10, the total absorptivity of CO_2 at the conditions given when exposed to 800 K black-body radiation. Repeat the calculation using Fig. 5-6 and Hottel's rule, Eq. (5-39).

6 RADIATION TRANSFER WITH AN ISOTHERMAL GAS

6.A Heat Transfer at a Black Wall. Often for design purposes a combustion chamber or chemical reactor can be modeled as containing a well-mixed gas, one with uniform temperature and composition, excluding the wall boundary layer. Suppose, for now, that the boundary layer has a negligible effect. The net spectral radiant heat flux across any plane can be written

$$q = \int_0^{2\pi}\int_0^{\pi/2} (I^+ - I^-)\cos\theta \sin\theta \, d\theta d\phi \tag{6-1}$$

At a black wall

$$I^+ = I_b(T_w) = B_w/\pi \tag{6-2}$$

and, from the solution to Eq. (5-6) with $I = I_b(T_w)$ at $s = 0$

$$I^- = I_b(T_w)e^{-k_a L(\theta,\phi)} + I_b(T_g)[1-e^{-k_a L(\theta,\phi)}] \tag{6-3}$$

Here the term $I_b(T_w)$ is the intensity of the wall across the chamber a distance $L(\theta,\phi)$ away, $e^{-k_a L(\theta,\phi)}$ is the transmissivity of the gas, and $[1-e^{-k_a L(\theta,\phi)}]$ is the emissivity of the gas. The irradiation is seen to be made up of two contributions, transmitted wall radiation and emitted gas radiation. Substitution into Eq. (6-1) gives, assuming the wall is isothermal,

$$q = (B_w - B_g)\frac{1}{\pi}\int_0^{2\pi}\int_0^{\pi/2} [1-e^{-k_a L(\theta,\phi)}]\cos\theta \sin\theta \, d\theta d\phi \tag{6-4}$$

The heat flow is

$$\dot{Q} = (B_w - B_g) \int_{A_w} \frac{1}{\pi} \int_0^{2\pi} \int_0^{\pi/2} [1 - e^{-k_a L(\theta,\phi)}] \cos\theta \sin\theta \, d\theta d\phi dA_w$$

which may be expressed as

$$\dot{Q} = (B_w - B_g) A_w \mathscr{F}_{wg} \tag{6-5}$$

where the wall-to-gas transfer factor is

$$\mathscr{F}_{wg} = \frac{1}{A_w} \int_{A_w} \frac{1}{\pi} \int_0^{2\pi} \int_0^{\pi/2} [1 - e^{-k_a L(\theta,\phi)}] \cos\theta \sin\theta \, d\theta d\phi dA_w \tag{6-6a}$$

Because, from Eq. (1-4),

$$d\Omega = \frac{\cos\theta' dA_w'}{L^2}$$

one can write equally well

$$\mathscr{F}_{wg} = \frac{1}{A_w} \int_{A_w} \int_{A_w'} [1 - e^{-k_a L}] \frac{\cos\theta \cos\theta'}{\pi L^2} dA_w' dA_w \tag{6-6b}$$

One finds that the transfer factor from a black isothermal wall to an isothermal gas is the shape-factor-weighted gas emissivity along all possible paths from the wall through the gas. Of course, Eq. (6-5) requires spectral integration. For now we focus upon the geometric aspects of the transfer process.

6.B <u>The Mean Beam Length Concept</u>. In view of Eq. (6-6b) it is natural to define a mean beam length L_{mb} such that

$$\mathscr{F}_{wg} = [1-e^{-k_a L_{mb}}]F_{w-w'} = [1-e^{-k_a L_{mb}}] \qquad (6\text{-}7)$$

where for a complete enclosure $F_{w-w'} = 1$. A strong motivation in defining L_{mb} is the expectation that an engineering approximation for it will be easily found.

Consider, as a specific example, the situation for the sphere as pictured in Fig. 6-1. There it is clear that $L(\theta,\phi) = 2R\cos\theta$. Thus Eq. (6-7) reduces to

$$[1-e^{-k_a L_{mb}}]_{sphere} = \int_0^{\pi/2} [1-e^{-2k_a R\cos\theta}]2\cos\theta \sin\theta \, d\theta \qquad (6\text{-}8)$$

The answer for $k_a L_{mb}$ is seen to be a function of the optical depth, say across the diameter, $k_a D$. It may be written formally

$$L_{mb}/D = \frac{1}{t} \ln \frac{t^2/2}{(1-e^{-t})-te^{-t}} \; , \quad t = k_a D \qquad (6\text{-}9)$$

but the form of Eq. (6-8) shows the situation more physically. The possible paths are seen to range in length from zero to one diameter. Clearly the mean beam length of the sphere will be less than the diameter.

In the limit as absorption coefficient goes to zero, that is, when the gas is optically thin, the approximation can be made

$$[1-e^{-k_a L_{mb}}] \doteq k_a L_{mb}$$

$$[1-e^{-k_a L}] \doteq k_a L$$

FIGURE 6-1

Path Length in a Sphere

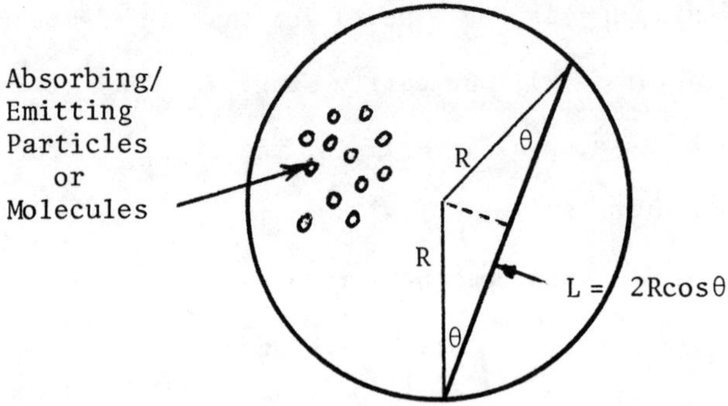

Then Eq. (6-8) with $\mu = \cos\theta$ reduces to

$$L_{mbg,sphere} = D\int_0^1 2\mu^2 d\mu = \frac{2}{3} D \qquad (6\text{-}10)$$

The subscript g here denotes the geometric mean beam length

$$L_{mbg} = \lim_{k_a \to 0} L_{mb} \qquad (6\text{-}11)$$

so called because the value of k_a cancels out, and only the geometry governs the value.

If one tries the approximation

$$\mathscr{F}_{wg} \equiv 1-e^{-k_a L_{mb}} \doteq 1-e^{-k_a L_{mbg}} \qquad (6\text{-}12)$$

and compares it with exact calculation of Eq. (6-8), one finds for the sphere that the approximate value of $\mathscr{F}_{wg}$ is within one per cent of the true value when $\kappa_a L_{mbg}$ is less than 0.25 and at worst is only 5.2 per cent high. If one repeats the comparison for the infinite cylinder and slab, one finds the accuracy is not as good, but if one uses $L_{mb} \doteq 0.9 L_{mbg}$ when $k_a L_{mbg}$ exceeds say 0.1, then the error is less than 7 per cent.

We see that an estimate of the geometric mean beam length leads easily to an engineering estimate of the transfer factor via Eq. (6-12). Thus it is of considerable interest to know that L_{mbg} is very readily found for the one-wall-one-gas enclosure. Recall that in finding a transfer factor one can think of the source black body radiosity, here B_g, as having the value unity and the sink, here $\dot{B}_w$, as having the value zero.

Using this concept and combining Eqs. (6-5), (6-7), and (6-11) gives

$$L_{mbg} = \lim_{k_a \to 0} \dot{Q}/A_w k_a B_g$$

But we have found that for an optically thin sphere

$$\dot{Q} = 4\pi R^2 B_g \frac{2}{3} Dk_a = \frac{16}{3} \pi R^3 k_a B_g$$

$$\dot{Q} = 4Vk_a B_g \ , \quad V = \frac{4}{3} \pi R^3$$

Any arbitrary volume of gas V_g can be filled with spheres with a spectrum of sizes as closely as desired. Since the volumes sum to V_g, the total power emitted sums to

$$\dot{Q} = 4V_g k_a B_g \tag{6-13}$$

Substituting into the expression for L_{mbg} gives

$$L_{mbg} = 4V_g/A_w \tag{6-14}$$

Equation (6-14) states that the geometric mean beam length is simply four times the gas volume divided by the wall area totally enclosing it. With this easily calculated value, the transfer factor is readily approximated via Eq. (6-12), and the heat transfer is found from Eq. (6-5). Note that for a long duct or tunnel where $A_w = P_w L$ and $V_g = A_c L$

the geometric mean beam length is identical to the hydraulic diameter $4A_c/P_w$.

The mean beam length concept, derived here on a spectral basis, can be put on a band or total basis. Integrate Eq. (6-4) or (6-5) over the spectrum. If the band approximation is suitable, one obtains

$$\dot{Q} = \sum_k (B_{w,k} - B_{g,k}) A_w \omega_k A_k^*(\rho_a L_{mb,k}) \tag{6-15}$$

where

$$A_k^*(\rho_a L_{mb,k}) = \frac{1}{A_w} \int_{A_w} \frac{1}{\pi} \int_0^{2\pi} \int_0^{\pi/2} A_k^*(\rho_a L) \cos\theta \sin\theta \, d\theta d\phi dA_w \tag{6-16}$$

In general, the exact value of the mean beam length is different for each of the absorption bands, but, in practice, one may approximate $L_{mb,k}$ with L_{mbg}, or a fraction of L_{mbg} appropriate to the most important band. This fraction has been found for the exponential band model with overlapped lines ($\eta = \beta P_e > 1$) for the sphere, cylinder, and slab geometries as shown in Table 6-1. The table shows that for a strong band $L_{mb} = 0.9L_{mbg}$ will do nicely for the cylinder or sphere, but for a slab $L_{mb} = 0.83L_{mbg}$.

TABLE 6-1

MEAN BEAM LENGTH FOR THE SLAB, CYLINDER, AND SPHERE

OPTICAL DEPTH AT BAND HEAD (based on plate spacing or diameter)	RATIO OF MEAN BEAM LENGTH TO GEOMETRIC MEAN BEAM LENGTH L_{mb}/L_{mbg}		
	SLAB	CYLINDER	SPHERE
0.00	1.00	1.00	1.00
0.01	0.99	1.00	1.00
0.02	0.98	1.00	1.00
0.05	0.96	1.00	1.00
0.10	0.95	1.00	1.00
0.20	0.93	1.00	1.00
0.50	0.88	0.98	0.99
1.0	0.86	0.95	0.98
2.0	0.84	0.93	0.97
5.0	0.83	0.91	0.94
10.0	0.83	0.90	0.92
∞	0.83	0.89	0.91

6.C <u>Wall Layer Transmission</u>. Section 7 deals with radiation transfer within a nonisothermal gas. At a cold black wall exposed to hot gases, the irradiation upon the wall is less when a cold boundary layer exists, because not all increments of path along an incident beam are at the high temperature. Based upon an analysis of thermally developing entrance flow of an absorbing-emitting molecular gas in a parallel-plate duct, Balakrishnan and Edwards [9] proposed that Eq. (6-15) be multiplied by a wall layer transmission factor $\tau_{WL,k}$

$$\dot{Q}_{0-x}/A_w = \sum_k (B_{w,k}-B_{g,k})\omega_k A_k^*(\rho_a L_{mb,k})\tau_{WL,k} \tag{6-17}$$

where

$$\tau_{WL,k} \doteq \frac{\ln(1+\tau_{HD,k})-\ln(1+\tau_{HD,k}r_{WL})}{\ln(1+\tau_{HD,k})} \tag{6-18a}$$

$$r_{WL} \doteq \frac{1.37}{Re^{0.37}}\left\{1 - \exp[-(x/10D)^{0.6}]\right\} \tag{6-18b}$$

where D is hydraulic diameter, Re is Reynolds number based upon hydraulic diameter, x is the length from the beginning of the developing thermal boundary layer, and $\tau_{HD,k}$ is the optical depth at the kth band head based upon the hydraulic diameter (identical to the geometric mean beam length), $\alpha_k \rho_a D/\omega_k$. Note that Eq. (6-17) is *not* for the local heat flux at a value of x, but the average heat flux for the region between 0 and x.

6.D Radiosity-Irradiation Formulation at an Isothermal-Gas-Filled-Enclosure Wall. On a true spectral basis Eqs. (6.2) and (6.3) may be replaced with

$$I_i^+ = \varepsilon_i I_b(T_i) + (1-\varepsilon_i) I_i^- \tag{6-19}$$

$$I_i^- = I_j^+ e^{-k_a L} + I_b(T_g)[1-e^{-k_a L}] \tag{6-20}$$

When these expressions are substituted into Eq. (6-1), and a mean beam length concept is invoked, the flux is

$$q_i = q_i^+ - q_i^- \tag{6-21}$$

$$q_i^+ = \varepsilon_i B_i + (1-\varepsilon_i) q_i^- \tag{6-22}$$

$$q_i^- = \sum_{j=1}^{n} F_{i-j}[q_j^+ \tau_{g,i-j} + (1-\tau_{g,i-j}) B_g] \tag{6-23}$$

where

$$\tau_{g,i-j} = e^{-k_a L_{i-j}} = \frac{1}{A_i F_{i-j}} \int_{A_i} \int_{A_j} e^{-k_a L} \frac{\cos\theta_i \cos\theta_j}{\pi L^2} \, dA_j \, dA_i \tag{6-24}$$

The path L is the distance between differential areas dA_j and dA_i, and L_{i-j} is the mean beam length for the finite areas A_j and A_i.

Oppenheim and Bevans [7] derived expressions for the geometric mean L_{i-j} for adjacent and opposite rectangles, and Dunkle [8] integrated the expressions and provided numerical values. Table 6.2 gives the integrated expressions

TABLE 6.2

CLOSED-FORM EXPRESSIONS FOR GEOMETRIC MEAN BEAM LENGTHS BETWEEN RECTANGLES
(See Table 3.1 for F_{12} expressions)

Geometry	Expression

1. Opposite Rectangles

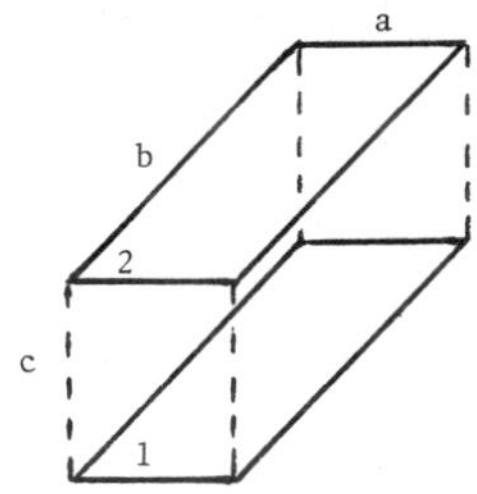

X = a/c
Y = b/c
$Z = L_{g,1-2}/c$

$$ZF_{1-2} = \frac{4}{\pi XY}\left\{ XY \tan^{-1}\frac{XY}{\sqrt{1+X^2+Y^2}} + X\,ln\,\frac{X+\sqrt{1+X^2+Y^2}}{(X+\sqrt{1+X^2})\sqrt{1+Y^2}} + Y\,ln\,\frac{Y+\sqrt{1+X^2+Y^2}}{(Y+\sqrt{1+Y^2})\sqrt{1+X^2}} + \sqrt{1+X^2} + \sqrt{1+Y^2} - \sqrt{1+X^2+Y^2} - 1 \right\}$$

2. Adjacent Rectangles

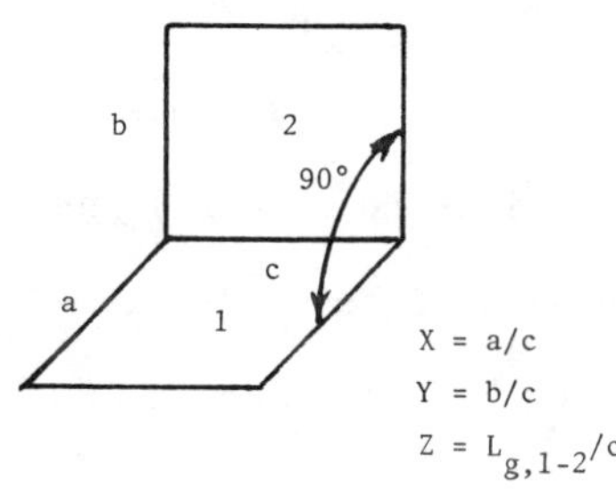

X = a/c
Y = b/c
$Z = L_{g,1-2}/c$

$$ZF_{1-2} = \frac{1}{3\pi X}\left\{ 3X^2\,ln\,\frac{(1+\sqrt{1+X^2})\sqrt{X^2+Y^2}}{X(1+\sqrt{1+X^2+Y^2})} + 3Y^2\,ln\,\frac{(1+\sqrt{1+Y^2})\sqrt{X^2+Y^2}}{Y(1+\sqrt{1+X^2+Y^2})} + 3X^2\left[\sqrt{1+X^2+Y^2} - \sqrt{X^2+Y^2} - \sqrt{1+X^2}\right] + 3Y^2\left[\sqrt{1+X^2+Y^2} - \sqrt{X^2+Y^2} - \sqrt{1+Y^2}\right] + (1+X^2)^{3/2} + (1+Y^2)^{3/2} + (X^2+Y^2)^{3/2} - (1+X^2+Y^2)^{3/2} + 2X^3 + 2Y^3 - 1 \right\}$$

and Table 6.3 some numerical values. Dunkle goes on to state that the $L_{g,1-2}$ for a finite sphere of radius R to a rectangle is the value for the point sphere minus (2/3)R. He also recommends approximating the human form as a sphere with (2/3)R = 0.26 m. The values of geometric mean beam length between areas i and j, $L_{g,i-j}$ is a close upper limit to the true mean beam length L_{i-j}.

The rules of shape factor algebra, illustrated in Figures 3.1 through 3.3, apply to geometric mean beam lengths when G is understood to be

$$G_{i-j} = G_{j-i} = L_{g,i-j} A_i F_{i-j} \qquad (6\text{-}25)$$

It is clear that Eqs. (6-22) and (6-23) are of the same form as Eqs. (3-13) and (3-14), and solution by matrix inversion (or reiteration) may be accomplished as explained in Section 3.B.

6.E <u>The Radiation Network with a Gas</u>. Just as a network interpretation exists for Eqs. (3-12) to (3-14), so does one exist for Eqs. (6-21) to (6-23). Since Eqs. (6-21) and (6-22) are exactly Eqs. (3-12) and (3-13), Eqs. (3-49) and (3-50) remain in force, showing that each surface has a surface resistance as depicted in Fig. 3-5. Equations (3-47) and (3-48) are based upon Eqs. (3-12) and (3-14) and must be replaced by relations following from Eqs. (6-21) and (6-23).

TABLE 6.3

A SHORT TABLE OF GEOMETRIC MEAN BEAM LENGTHS

1. Opposite Rectangles

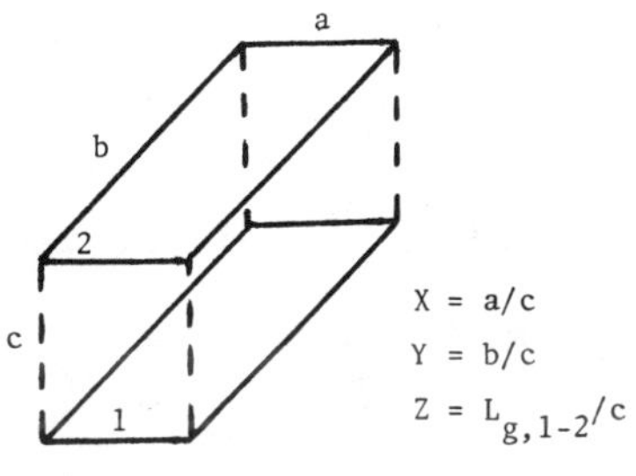

$X = a/c$
$Y = b/c$
$Z = L_{g,1-2}/c$

Z

	Y= 0.0	0.1	1	10	∞
X=0	1.00	1.00	1.05	1.23	1.27
0.1		1.00	1.06	1.23	1.27
1			1.11	1.30	1.35
10				1.62	1.75
∞					2.00

2. Adjacent Rectangles

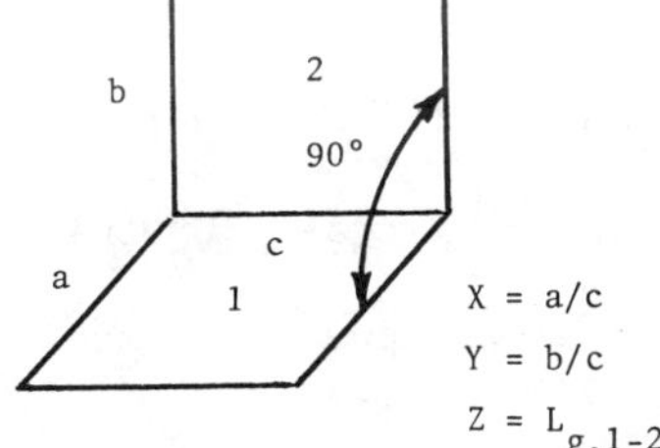

$X = a/c$
$Y = b/c$
$Z = L_{g,1-2}/c$

Z

	Y= 0.1	0.2	1	2	10	20
X=0.1	0.072	0.098	0.15	0.17	0.18	0.18
0.2		0.14	0.24	0.27	0.30	0.30
1			0.56	0.70	0.89	0.92
2				0.95	1.38	1.47
10					3.20	3.96
20						5.49

3. Point-Sphere to Rectangles

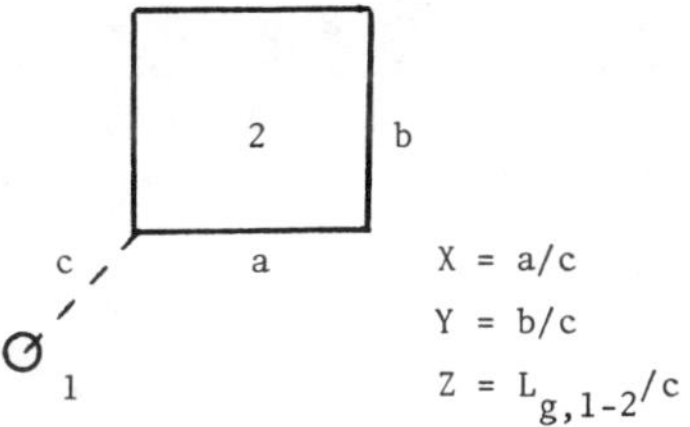

$X = a/c$
$Y = b/c$
$Z = L_{g,1-2}/c$

Z

	Y= 0	1	2	3
X=0	1.00	1.11	1.25	1.32
1		1.22	1.37	1.45
2			1.52	1.63
3				1.76

$$\dot{Q}_i = q_i A_i = \sum_{j=1}^{n} A_i F_{i-j} \tau_{g,i-j} (q_i^+ - q_j^+)$$

$$+ \sum_{j=1}^{n} A_i F_{i-j} (1 - \tau_{g,i-j})(q_i^+ - B_g) \qquad (6\text{-}26)$$

This relation shows (1) that the internodal resistances R_{i-j} become

$$R_{i-j} = \frac{1}{A_i F_{i-j} \tau_{g,i-j}} \qquad (6\text{-}27)$$

and (2) that for every internodal resistance R_{i-j} (including R_{i-i}) there is a node-to-gas-node resistance in parallel. Since parallel conductances add

$$\frac{1}{R_{i-g}} = \sum_{j=1}^{n} A_i F_{i-j} (1 - \tau_{g,i-j}) \qquad (6\text{-}28)$$

It is imperative that the F_{i-i} contribution (if any) not be overlooked. As an aid to the memory in this regard, Oppenheim advocated drawing the otherwise useless appendage $R_{i-i} = 1/A_i F_{i-i}$ on the network as shown in Fig. 6-2, which summarizes the rules for drawing a radiation network. Figure 6-3 shows a complete network for the case n = 2.

FIGURE 6-2

Radiation Network Connections for Node i with Isothermal Gas g

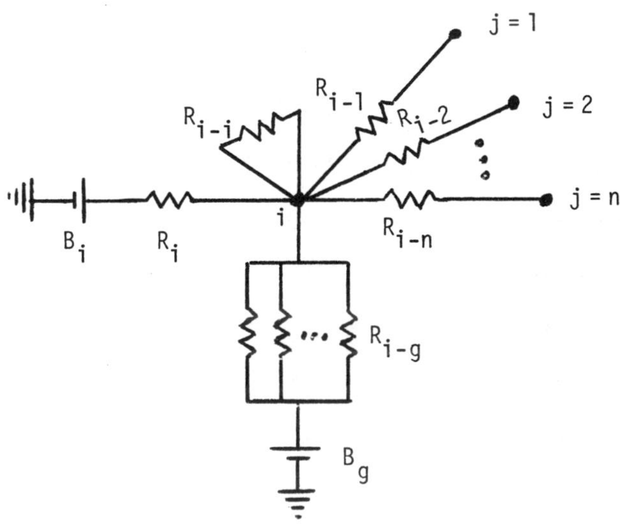

FIGURE 6-3

Radiation Network for n = 2 with Isothermal Gas g

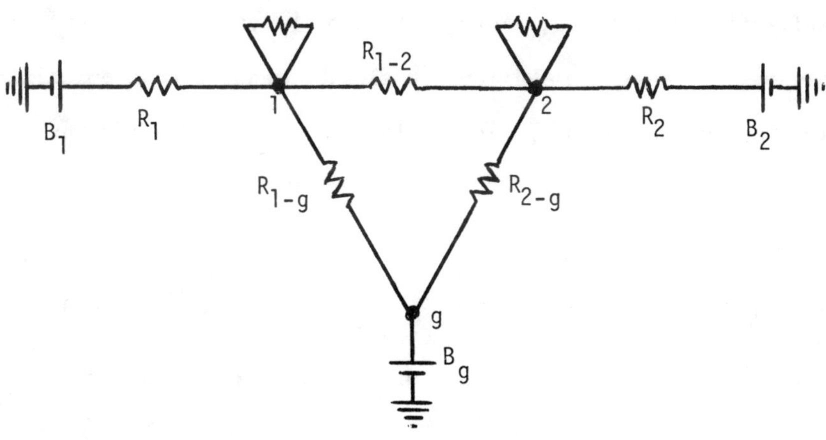

6.F <u>Some Useful Results</u>. For a gas containing spray droplets or particles, the spectral variation in absorption coefficient may often be overlooked. A gas like H_2O vapor with several wide weak bands may be similarly treated to a lower level of approximation. For the particle-laden gas, $\tau_{g,i-j}$ is $e^{-k_a L_{i-j}}$ as given by Eq. (6-24), and $\varepsilon_{g,i-j}$ is $1-\tau_{g,i-j}$. For the H_2O vapor $\varepsilon_{g,i-j}$ is found from Fig. 5-4 using the mean beam length L_{i-j} and $\tau_{g,i-j} = 1-\varepsilon_{g,i-j}$. Using the radiation network, one can quickly develop working relations for the following cases:

1. A gaseous source and a sink.
2. A gaseous source, a refractory, and a sink.
3. A source, a sink, and an adiabatic gas.
4. A source, a sink, a refractory, and an adiabatic gas.
5. A source, a sink, and a gaseous source or sink.

As in Section 3.E, the results are subject to assumptions of perfectly diffuse gray surfaces and, in addition, to well-stirred gray gas. Note that when the gas is a source or sink, it is presumably due to flow, perhaps with chemical reaction or change of phase, so that the heat flow from the gas $\dot{Q}_g$ is provided (or taken up) by an enthalpy change as follows

$$\dot{m}_g(\hat{h}_{g,i}-\hat{h}_{g,o}) = \dot{Q}_g \tag{6-29}$$

where $\dot{m}_g$ is the mass flow rate of the gas mixture through the enclosure, $\hat{h}_{g,i}$ is the specific enthalpy of the entering gas mixture including heats of formation or combustion, and $\hat{h}_{g,o}$ is the specific enthalpy of the exiting gas mixture at temperature T_g. As an approximation it is frequently possible to write Eq. (6-29) in the form

$$\dot{m}_g c_{p,g}(T_{gf}-T_g) = \dot{Q}_g \tag{6-30}$$

where T_{gf} is a fictitious gas temperature (not the completion-limited adiabatic flame temperature, it is emphasized) found from

$$T_{gf} = \frac{\dot{m}_g c_{p,f} T_{f,i} + \dot{m}_f \Delta\hat{h} + \dot{m}_a c_{p,a} T_a}{\dot{m}_g c_{p,g}} \tag{6-31}$$

where $\dot{m}_f$ is the fuel flow rate, $c_{p,f}$ its specific heat, $\Delta\hat{h}$ is the heat of reaction based on $\dot{m}_f$, $\dot{m}_a$ is the air flow rate, $c_{p,a}$ its specific heat, $\dot{m}_g$ is the gas product flow rate ($\dot{m}_g = \dot{m}_f + \dot{m}_a$), and $c_{p,g}$ is its specific heat. The point is that, while in what follows we assume we know T_g and solve for $\dot{Q}_g$, in fact we usually do not know T_g but instead $\hat{h}_{g,i}$ or T_{gf} and have to use Eq. (6-29) or (6-30) together with $\dot{Q}_g$ as a function of T_g to solve for T_g.

A gaseous source and a sink model a combustion chamber or reactor with a cooled wall. With $F_{w-w} = 1$, the network is simply a surface resistance and gas-to-wall resistance. Hence

$$A_w \mathscr{F}_{w-g} = \frac{\dot{Q}_g}{\sigma T_g^4 - \sigma T_w^4} = \frac{A_w}{\dfrac{1-\varepsilon_w}{\varepsilon_w} + \dfrac{1}{\varepsilon_g}}$$

A gaseous source, a refractory, and a sink models a direct-fired furnace. Figure 6-2 applies with battery B_2 open circuit to make the refractory wall 2 radiatively adiabatic.

$$A_1\mathscr{F}_{1-g} = \frac{1}{\frac{1-\varepsilon_1}{\varepsilon_1 A_1} + \frac{1}{\frac{1}{R_{1-g}} + \frac{1}{R_{1-2}+R_{2-g}}}} \qquad (6\text{-}33a)$$

where

$$\frac{1}{R_{1-g}} = A_1F_{1-1}\varepsilon_{g,1-1}+A_1F_{1-2}\varepsilon_{g,1-2} \qquad (6\text{-}33b)$$

$$R_{1-2} = \frac{1}{A_1F_{1-2}\tau_{g,1-2}} \qquad (6\text{-}33c)$$

$$\frac{1}{R_{2-g}} = A_1F_{1-2}\varepsilon_{g,1-2}+A_2F_{2-2}\varepsilon_{g,2-2} \qquad (6\text{-}33d)$$

If one needs to know the temperature of the refractory, one can solve for B_2 as follows:

$$q_1^+ = B_1+A_1\mathscr{F}_{1-g}(B_g-B_1)(1-\varepsilon_1)/\varepsilon_1A_1 \qquad (6\text{-}34a)$$

$$\dot{Q}_{1-2-g} = \frac{B_g-q_1^+}{R_{1-2}+R_{2-g}} \qquad (6\text{-}34b)$$

$$B_2 = q_2^+ = B_g-\dot{Q}_{1-2-g}R_{2-g} \qquad (6\text{-}34c)$$

$$T_2 = (B_2/\sigma)^{1/4} \qquad (6\text{-}34d)$$

A source, a sink, and an adiabatic gas may model a nuclear accident scenario where hot fuel rods radiate to cold surface (spacers, cannister walls, etc.) through dry stagnant steam. Figure 6-2 again applies with battery B_g open circuit to place the gas in radiative equilibrium. Then

$$A_1\mathcal{F}_{1-2} = \cfrac{1}{\cfrac{1-\varepsilon_1}{\varepsilon_1 A_1} + \cfrac{1}{\cfrac{1}{R_{1-2}} + \cfrac{1}{R_{1-g}+R_{2-g}}} + \cfrac{1-\varepsilon_2}{\varepsilon_2 A_2}} \tag{6-35}$$

where R_{1-2}, R_{1-g}, and R_{2-g} are as shown in Eqs. (6-33b - d).

A source, a sink, a refractory wall, and an adiabatic gas may be used to evaluate the effect of a dusty or otherwise absorbing atmosphere on the performance of an electric or muffle furnace, for example. Figure 6-3 applies with batteries B_3 and B_g open circuit. Although the network is more intricate, it remains a two-node one; only B_1 and B_2 are connected. One proceeds to simplify the network by converting the R_{1-3}, R_{2-3}, R_{3-g} wye into a delta by means of the wye-delta transformation in Fig. 3-7. Denote the delta resistances $R_{1-2,\Delta}$, $R_{1-g,\Delta}$ and $R_{2-g,\Delta}$. Then adding the conductances in parallel gives

$$\frac{1}{R_{1-2(3)}} = \frac{1}{R_{1-2}} + \frac{1}{R_{1-2,\Delta}} \tag{6-36a}$$

$$\frac{1}{R_{1-g(3)}} = \frac{1}{R_{1-g}} + \frac{1}{R_{1-g,\Delta}} \tag{6-36b}$$

$$\frac{1}{R_{2-g(3)}} = \frac{1}{R_{2-g}} + \frac{1}{R_{2-g,\Delta}} \tag{6-36c}$$

These resistances are entered into Eq. (6-35) to find the desired transfer factor.

FIGURE 6-4

Radiation Network for n = 3
with Isothermal Gas g

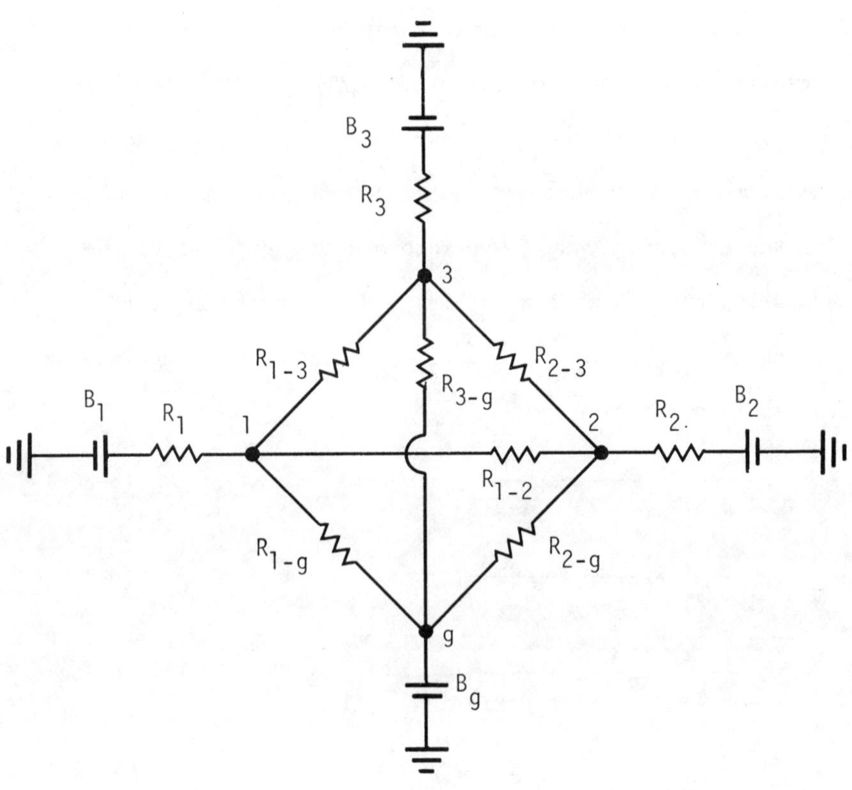

A source, a sink, and a gaseous source or sink may model a direct-fired furnace where the refractory is made nonadiabatic by convection, or a nuclear accident scenario when spray cooling droplets are evaporating, for example. Note that an additional truly adiabatic surface may be included by using $R_{1-2(3)}$, $R_{1-g(3)}$, and $R_{2-g(3)}$ given by Eqs. (6-36a - c) in place of R_{1-2}, R_{1-g}, and R_{2-g} shown on Fig. 6-3. With or without the addition of a truly adiabatic surface, the network is no longer a simple 2-node one, because three nodes are nonadiabatic. One proceeds by converting the 1-2-g delta in Fig. 6-3 into a wye according to Fig. 3-7. Denote the result as $R_{1,Y}$, $R_{2,Y}$, and $R_{g,Y}$. Then the resistances in series are added to give total values in each branch of a larger wye

$$R_{1(T)} = R_1+R_{1,Y} \ , \quad R_{2(T)} = R_2+R_{2,Y} \ , \quad R_{g(T)} = R_{g,Y} \tag{6-37}$$

Finally the wye is converted back to a delta according to Fig. 3-7.

$$A_1\mathcal{F}_{1-2} = \frac{R^{-1}_{1(T)}R^{-1}_{2(T)}}{R^{-1}_{1(T)}+R^{-1}_{2(T)}+R^{-1}_{g(T)}} \tag{6-38a}$$

$$A_1\mathcal{F}_{1-g} = \frac{R^{-1}_{1(T)}R^{-1}_{g(T)}}{R^{-1}_{1(T)}+R^{-1}_{2(T)}+R^{-1}_{g(T)}} \tag{6-38b}$$

$$A_2\mathcal{F}_{2-g} = \frac{R^{-1}_{2(T)}R^{-1}_{g(T)}}{R^{-1}_{1(T)}+R^{-1}_{2(T)}+R^{-1}_{g(T)}} \tag{6-38c}$$

6.G <u>Sample Calculation</u>. Suppose a 6 meter diameter furnace is fired by 0.15 kg/s of gaseous fuel having a lower heating value of $5x10^7$ J/kg and 2.7 kg/s of air, both air and fuel entering at 500 K. The load is at 900 K covered by slag 6 mm thick having a thermal conductivity of 2 W/m K and an emissivity of 0.48. The refractory roof is 50 m^2 in area. The furnace gas has a specific heat of 1200 J/kg K and, for the calculated mean beam length and at an estimated gas temperature, an emissivity $\varepsilon_g = 0.25$. It is desired to calculate T_g, T_1, and the heat transfer rate into the load.

If we neglect convection, we have case 2, a gaseous source, a refractory, and a sink. We begin by calculating $A_1\mathcal{F}_{1-g}$ from Eq. (6-33). The following values pertain (assuming $\varepsilon_{g,1-2} \doteq \varepsilon_{g,2-2}$)

$$R_1 = \frac{1-\varepsilon_1}{\varepsilon_1 A_1} = \frac{1-0.48}{(0.48)(28.3)} = 3.83x10^{-2}\ m^{-2}$$

$$R_{1-2} = \frac{1}{A_1 F_{1-2} \tau_{g,1-2}} = \frac{1}{(28.3)(1)(0.75)} = 4.72x10^{-2}\ m^{-2}$$

$$R_{1-g} = \frac{1}{A_1 \varepsilon_{g,1-2}} = \frac{1}{(28.3)(0.25)} = 14.15x10^{-2}\ m^{-2}$$

$$R_{2-g} = \frac{1}{A_2 F_{2-1} \varepsilon_{g,2-1} + A_2 F_{2-2} \varepsilon_{g,2-2}} = \frac{1}{(50)(0.25)}$$

$$= 8.00x10^{-2}\ m^{-2}$$

$$A_1\mathcal{F}_{1-g} = \frac{1}{3.83x10^{-2} + \cfrac{1}{\cfrac{1}{14.15x10^{-2}} + \cfrac{1}{(4.72+8.00)x10^{-2}}}} = 9.5\ m^2$$

Next, Eq. (6-30) is written for the heat balance on the gas

$$\dot{m}_g c_{p,g}(T_{gf}-T_g) = A_1\mathcal{F}_{1-g}(\sigma T_g^{\ 4}-\sigma T_1^{\ 4}) \qquad (6\text{-}39)$$

where

$$\dot{m}c_{p,g} = (2.85)(1200) = 3420 \text{ W/K}$$

$$T_{gf} = 500+0.15(5\text{x}10^7)/3420 = 2693 \text{ K}$$

A heat balance on the surface between the s and m interfaces is written for area A_1.

$$A_1\mathcal{F}_{1-g}(\sigma T_g^{\ 4}-\sigma T_1^{\ 4}) = \frac{kA_1}{\delta}(T_1-T_o) \qquad (6\text{-}40)$$

where

$$\frac{kA_1}{\delta} = \frac{(2)(28.3)}{6\text{x}10^{-3}} = 9425 \text{ W/K}$$

The two nonlinear equations must be solved simultaneously for T_g and T_1. In a hand calculation one assumes an independent value $\dot{Q}_{g-1}^{(i)}$ for the heat transfer rate

$$\dot{Q}_{g-1} = A_1\mathcal{F}_{1-g}(\sigma T_g^{\ 4}-\sigma T_1^{\ 4}) \qquad (6\text{-}41)$$

enters it into the right-hand side of Eq. (6-39) and solves for T_g and then into the left-hand side of Eq. (6-40) and solves for T_1. The values of T_g and T_1 are used to find a dependent value of $\dot{Q}_{g-1}^{(d)}$ from Eq. (6-41). Correct values of $\dot{Q}_{g-1}$, T_1, and T_g are obtained when $\dot{Q}_{g-1}^{(i)} = \dot{Q}_{g-1}^{(d)}$. The following table completes the example.

$\dot{Q}^{(i)}_{g-1}$	T_g	T_1	$\dot{Q}^{(d)}_{g-1}$
W	K	K	W
$3x10^6$	1816	1218	$4.67x10^6$
$4x10^6$	1523	1324	$1.24x10^6$
$3.3x10^6$	1728	1250	$3.49x10^6$
$3.34x10^6$	1716	1254	$3.34x10^6$

The answers are seen to be $\dot{Q}_{g-1} = 3.34x10^6$ W, $T_g = 1716$ K and $T_1 = 1254$ K. Having calculated a value of T_g, one should redetermine the value of ε_g and repeat the calculation.

If we considered convective coupling, we would have to use Case 5, a source, a sink, and a gaseous source. Then $A_1\mathcal{F}_{1-2}$, $A_1\mathcal{F}_{1-g}$, and $A_2\mathcal{F}_{2-g}$ would be found, and one would have three simultaneous nonlinear equations to solve for T_1, T_2, and T_g.

$$\dot{m}_g c_{p,g}(T_f-T_g) = h_1A_1(T_g-T_1)+h_2A_2(T_g-T_2)+A_1\mathcal{F}_{1-g}(\sigma T_g^{\ 4}-\sigma T_1^{\ 4}) + A_2\mathcal{F}_{2-g}(\sigma T_g^{\ 4}-\sigma T_2^{\ 4}) \tag{6-42}$$

$$h_2A_2(T_g-T_2)+A_2\mathcal{F}_{2-g}(\sigma T_g^{\ 4}-\sigma T_2^{\ 4}) = A_1\mathcal{F}_{1-2}(\sigma T_2^{\ 4}-\sigma T_1^{\ 4}) \tag{6-43}$$

$$h_1A_1(T_g-T_1)+A_1\mathcal{F}_{1-g}(\sigma T_g^{\ 4}-\sigma T_1^{\ 4})+A_1\mathcal{F}_{1-2}(\sigma T_2^{\ 4}-\sigma T_1^{\ 4}) = \frac{kA_1}{\delta}(T_1-T_0) \tag{6-44}$$

6.H Monte Carlo Solution. How is the working engineer going to solve a more general radiation heat transfer problem than those five cases enumerated in Section 6.F. A general method easy to grasp is the Monte Carlo one [10,11]. Suppose the temperature and composition field in the medium is known well enough to fix the properties. Divide the medium into M subvolumes and the surface into N areas. If only the heat flows for the areas are required, one need find only the transfer factor from area to area and area to volume. If one needs to know the heat flows for the volumes also, one will have to find the transfer factors from volume to volume as well. To find a transfer factor, one imagines that only one source area or volume i is turned on, $B_i = \pi I_{bi} = 1$, and all the others are off. Then in the Monte Carlo sense one chooses randomly a source point and a direction of emission for the ray to be traced. For a surface, the ray is weighted by its directional emissivity and $dA \cos\theta \, d\Omega$ as explained previously. For a volume of isotropic emitting material the weighting is $k_a dV d\Omega$; that is, the points chosen within the volume should be uniformly distributed over the volume, and the direction should be uniform in solid angle. Thus given two random numbers P_1 and P_2, the direction with respect to an arbitrary polar-azimuthal coordinate system would be

$$\theta = \cos^{-1}(2P_1 - 1) \qquad (6\text{-}45)$$

$$\phi = 2\pi P_2 \tag{6-46}$$

since $d\Omega = \sin\theta \, d\theta \, d\phi = -d(\cos\theta)d\phi$.

The ray tracing proceeds as described in Section 4, except that it is necessary to move along a ray in steps of Δs instead of going the entire distance R from surface to surface. The increment Δs is taken so that there are, say, an integer number of steps across the jth volume element being traversed. With each step Δs the probability of absorption or scattering is compared to a random number P. From Eq. (5-8) this probability is $(1-e^{-k_e\Delta s})$. If P is greater than this probability, the ray continues along its original direction unchanged. If P is less than $\omega_s(1-e^{-k_e s})$ the photon bundle is scattered. If it is not transmitted or scattered, it is absorbed. If the ray is scattered, a new direction is chosen according to Eqs. (6-45) and (6-46) (the ray energy is weighted by the phase function p if anisotropic scattering is considered), and the ray tracing continues until the requisite number of absorptions are scored.

Suppose that, in a Monte Carlo calculation of 10000 rays from surface i (weighted by say directional emissivity divided by hemispherical emissivity), 3000 were absorbed by surface or volume j. Call f_{ij} the fraction 3000/10000. Then the heat flow transfer factor is

$$\mathscr{G}_{ij} = A_i \mathscr{F}_{ij} = \varepsilon_i A_i f_{ij} \tag{6-47}$$

Suppose that, of 10000 rays sent out from volume i, 2000 were absorbed by surface or volume j. Call g_{ij} the fraction 2000/10000. The surface emission is weighted by $\varepsilon_i A_i$, because the emitted heat flux is $\varepsilon_i B_i$, and the emitted heat flow is the flux times the area. An equivalent weighting is necessary for a volume.

Recall Eq. (6-13). The emitted radiation per unit volume is

$$\dot{Q}_{V,emitted} = \dot{Q}_{emitted}/V_g = 4k_a B_g \tag{6-48}$$

Thus the emission from the ith volume element is $4k_{a,i}V_iB_i$ and of that emission a fraction g_{ij} is absorbed by element j. Hence the heat flow transfer factor is

$$\mathscr{G}_{ij} = 4k_{a,i}V_i g_{ij} \tag{6-49}$$

We need not be concerned that in a large volume some of the emitted radiation is scattered or reabsorbed. That fact was accounted for in the Monte Carlo calculation of g_{ii} and g_{ij}.

The notation adopted here for $\mathscr{G}_{ij}$ does not distinguish between a volume and a surface. After all, a "surface" is a volume, the volume between the s and m thermodynamic control surfaces (where k_a is large). Because a photon can traverse a path from i to j equally well from j to i, reciprocity holds

$$\mathscr{G}_{ij} = \mathscr{G}_{ji} \tag{6-50}$$

regardless of whether i and j are both surfaces, or one or both are volumes.

With knowledge that the heat flow emitted by i and absorbed by j is $B_i \mathscr{G}_{ij}$, one can find the *net* heat flow between i and j

$$\dot{Q}_{ij} = \mathscr{G}_{ij}(B_i - B_j)$$

If temperatures T_i and T_j are equal so $B_i = B_j$, then $\dot{Q}_{ij} = 0$. The total heat flow out of element i is the sum

$$\dot{Q}_i = \sum_j \mathscr{G}_{ij}(B_i - B_j) \qquad (6\text{-}51)$$

Note that the heat flux factor $\mathscr{G}_{ii}$ is of no consequence. However, in addition to Eq. (6-51), a good check for a Monte Carlo calculation is to sum $\mathscr{G}_{ij}$ over all j including j = i. Since the sum of f_{ij} or g_{ij} over all j is unity, the sum of $\mathscr{G}_{ij}$ over all N surfaces and M volumes is

$$\sum_{j=1}^{N+M} \mathscr{G}_{ij} = \begin{cases} \varepsilon_i A_i & i=1,\ldots,N \\ 4k_{a,i} V_i & i=n+1,\ldots,N+M \end{cases} \qquad (6\text{-}52)$$

A zone method akin to the Monte Carlo method is to subdivide the volume into M zones and the surface into N subsurfaces just as for the Monte Carlo method, but instead of computing the transfer factor directly, the radiosity-irradiation problem is formulated. In the absence of scattering the procedure is relatively straightforward, but tedious if more than one gas volume is involved, because shape factors involving gas transmission must be computed. Indeed, one way of computing such factors is to employ the Monte Carlo concept, so it is difficult to see why the transfer factor should not be computed directly via that algorithm. With scattering, the volume-to-volume and volume-to-surface shape factors are computed much as described in the Monte Carlo method, but scattered rays need not be traced, giving some computational advantage. One considers only the straight-line paths and scores rays either scattered or absorbed. The radiosity formulation is then broadened to include the source function S_i of each of the M volumes as well as the radiosity q_i^+ of the N surfaces. Just as the product of the shape factor F_{i-j} and reflectivity ρ_i gives the contribution to the radiosity of surface i by the radiosity at surface j for transfer in a diathermanous medium, so the products of the albedos and computed factors, call them $\omega_i F^{**}_{i-j}$, gives the contribution to the source function at volume i by the radiosity q_j^+ (for a surface) or source function S_j (for a volume). Matrix inversion or iterative solution then gives the set of S_i (i = 1,M) and q_i^+ (i = m+1, M+N).

REFERENCES TO SECTION 6

I. The Engineering Enclosure Problem

1. Hottel, H. C., "Radiant Heat Transmission", Chapter 4 of W. H. McAdams' Heat Transmission, McGraw-Hill, New York, 3rd edition, 1954. See also Hottel, H.C., and A. F. Sarofim, Radiative Transfer, McGraw-Hill, New York, 1967.

2. Oppenheim, A. K., "The Engineering Radiation Problem--An Example of the Interaction Between Engineering and Mathematics", Zeitschrift fur angewandte Mathematik und Mechanik, Vol. 36, pp. 81-93, 1956.

3. Bevans, J. T., and R. V. Dunkle, "Radiant Interchange within an Enclosure", J. Heat Transfer, Vol. 82, pp. 1-19, 1960.

4. Edwards, D. K., "Molecular Gas Band Radiation", Advances in Heat Transfer, Vol. 12, pp. 115-193, 1976.

5. Becker, H. B., "A Mathematical Solution for Gas-to-Surface Radiative Exchange Area for a Rectangular Parallelepiped Enclosure Containing a Gray Medium", J. Heat Transfer, Vol. 99, pp. 203-211, 1977.

6. Nelson, D. A., "Band Radiation within Diffuse-walled Enclosures", J. Heat Transfer, Vol. 101, pp. 81-89, 1979.

II. Mean Beam Lengths (see also Refs. [1] and [4])

7. Oppenheim, A. K., and J. T. Bevans, "Geometric Factors for Radiant Heat Transfer Through an Absorbing Medium in Cartesian Coordinates", J. Heat Transfer, Vol. 82, pp. 360-368, 1960.

8. Dunkle, R. V., "Geometric Mean Beam Lengths for Radiant Heat Transfer Calculations", J. Heat Transfer, Vol. 86, pp. 75-80, 1964.

III. Wall Layer Transmission

9. Balakrishnan, A., and D. K. Edwards, "Molecular Gas Radiation in the Thermal Entrance Region of a Duct", J. Heat Transfer, Vol. 101, pp. 489-495, 1979.

IV. Monte Carlo

10. Howell, J. R., and M. Perlmutter, "Monte Carlo Solution of Thermal Transfer through Radiant Media Between Gray Walls", J. Heat Transfer, Vol. 86, pp. 116-122, 1964.

11. Stockham, L. W., and T. J. Love, "Radiative Heat Transfer from a Cylindrical Cloud of Particles", AIAA Journal, Vol. 6, pp. 1935-1940, 1968.

EXERCISES FOR SECTION 6

1. (a) Find the ratio of the mean beam length L_{mb} to the geometric mean beam length L_{mbg} for the sphere geometry. Take monochromatic values of $k_a L_{mbg} = 0.01, 0.1, 1.0, 10, 100$. (b) For $B_g = 1$ and $B_w = 0$ and a black-walled sphere find the error in the wall heat flux $e = (q_{approx} - q_{exact})/q_{exact}$ from using $L_{mb,approx} = 0.9L_{mbg}$. Use the above values of $k_a L_{mbg}$.

2. Read Section 7.C and note Table 7-1. Then repeat Exercise 1 for the slab geometry. Discuss briefly the differences in the results.

3. Find the ratio of mean beam length to geometric mean beam length for a spherical black isothermal enclosure surrounding an isothermal gas with a single narrow absorption band having a band absorption of $A = C_2[P_e X]^{\frac{1}{2}}$ where C_2 is a constant and X is the absorber-density-beam-length product.

4. Repeat problem 3 for $A = \omega[ln\tau_H + E_1(\tau_H) + \gamma]$ where $\tau_H = \alpha X/\omega$. Verify the last column in Table 6-1.

5. A 1400°C gray isothermal gas with $k_a = 0.25\ m^{-1}$ is contained within a 4 m diameter 450°C sphere. (a) Find the wall heat flux when the wall is black. (b) Find the wall

5. (continued)

heat flux when the wall has an emissivity of 0.6. (c) Why isn't the answer to (b) 0.6 times the answer to (a)? Find your answers to parts (b) and (c) using first the radiosity and irradiation equations. Then use the network method. Compare the ease of using the methods.

6. A mixture of carbon-dioxide and nitrogen gases are at 1000 K and 1 atm total pressure in a large char-fired second-stage combustor with mean beam length L = 5 m and $P_a L = 1$ atm-m. The black walls of the combustor are at 800 K. What is the radiative heat flux at the wall?

7. A direct-fired furnace is in the form of a short vertical cylinder 6 m in diameter and 2 m high with an ellipsoidal roof, the roof having an area of 40 m^2. The side walls and roof are refractory, the gas is at 1400°C and has an effective gray absorption coefficient of 0.4 m^{-1}. The 900°C load covers the floor and has an emissivity of 0.55. (a) Find the heat flow into the load. (b) How much could the heat transfer rate be improved by raising the load emissivity to 0.9?

8. Municipal refuse is radiantly heated in order to generate gaseous fuel by pyrolysis. The refuse is conveyed along the floor of a long tunnel kiln 2 m wide by 1 m high. The ceiling and side walls of the tunnel are

8. (continued)

muffle heated to temperature T_1 in the pyrolysis section where the refuse decomposes endothermically at $T_2 = 500$ K. The gas product fills the tunnel and is estimated to have an effective gray absorption coefficient of 0.5 m^{-1}. (a) If it is assumed that the gas is well mixed at a temperature T_g dictated by radiative equilibrium, and T_g is not to exceed 800 K, what is the maximum value T_1 can have? Treat the refuse as black and the tunnel surfaces as gray with $\varepsilon_1 = 0.55$. (b) Allow for the flow of gas in at T_2. Take $c_p = 1250$ J/kg K.

9. In a hypothetical nuclear reactor loss-of-coolant accident, emergency spray cooling is used to remove decay heat. It is desired to estimate the radiative cooling to droplets below the quench front. Fuel pins 12 mm in diameter with a surface temperature of 1500 K and an emissivity of 0.7 are in a square array with a center-to-center pitch of 15 mm. In the space between the rods is wet steam containing 100 μm droplets. The steam has a quality of 2 per cent and is at 420 K. (a) Find the geometric mean beam length for the space between rods. (b) Assuming $Q_{abs} = 1$ find the absorption coefficient of the droplets. (c) Find the radiative heat flux at the surface of the rods. (d) Find the instantaneous time-rate-of-change of the steam quality under the conditions stated.

10. A direct-fired furnace is cubical in shape 3 m on a side. The walls and ceiling are refractory, and the load at 1000 K covers the floor and has an emissivity of 0.70. Furnace gases with a specific heat of 1200 J/kg K enter at T_{gf} = 2200 K and mix with the furnace gas at temperature T_g. The furnace gas then flows to the stack at that temperature. The effectiveness of the furnace is defined as $\varepsilon = \dot{Q}_{load}/\dot{m}c_p(T_{gf}-T_{load}) \doteq (T_{gf}-T_g)/(T_{gf}-T_{load})$ assuming constant c_p. Find the effectiveness when the furnace is fired at a rate sufficient to force $\dot{Q}_{load}$ = 3000 kW. Treat the gas as gray with an effective absorption coefficient of 0.20 m^{-1}.

7 NONISOTHERMAL GAS RADIATION

7.A Solution of the Equation of Transfer. Recall that Eq. (5-6) for the change of intensity along slant path increment ds,

$$\frac{dI}{ds} = -k_a I + k_a I_b \qquad (k_s = 0) \quad , \tag{7-1}$$

had a solution for the homogeneous isothermal gas given by Eq. (6-3),

$$I = I_o e^{-k_a s} + I_b(T_g)[1-e^{-k_a s}] \quad (I_b = \text{constant}) \tag{7-2}$$

Here $e^{-k_a s}$ is seen to be the transmissivity of a path s long, and $[1-e^{-k_a s}]$ is the emissivity. When k_a varies with distance s, for example if the composition varies, one merely transforms from distance s to optical depth t

$$t = \int_0^s k_a ds \tag{7-3}$$

and obtains in place of Eq. (7-1)

$$\frac{dI}{dt} = -I + I_b \tag{7-4}$$

The transmissivity of the nonhomogeneous gas is accordingly e^{-t}, and the emissivity is $1-e^{-t}$. The transformation simply says that the radiation doesn't care whether the radiating species is dilutely distributed over a large path or swept up into a short one. Meteorologists frequently describe the amount of water vapor in a path by giving the value of "precipital centimeters" of liquid water (actually the value of

$X = \rho_a L$ in g/cm^2, since $\rho_{liq} = 1$ g/cm^3). Of course, we know that the collision broadening as measured by the effect of P_e (recall Eq. (5-57)) on k_a would be affected if a molecular gas were swept up into a short path, but one considers the value of P_e as unchanged in transforming from s to t.

When T_g and hence $I_b(T_g)$ varies along the path, one finds the solution to Eq. (7-4) in a formal mathematical manner by variation of parameters. However, the engineer can picture the formal mathematical solution, and easily remember how to construct it, from the following physical considerations. Take an increment of path ds' at distance s' (optical depth t') along the path. According to Eq. (7-2) that element emits $I_b(T_g(s'))k_a ds'$ because $1-e^{-dt'} = dt' = k_a ds'$ is the emissivity of the increment. The fraction transmitted from optical depth t' to t is $e^{-(t-t')}$. Thus the contribution from increment ds' to the intensity at s is

$$dI = e^{-(t-t')} I_b(t')dt'$$

$$dI = e^{-\int_{s'}^{s} k_a(s'')ds''} I_b(s')k_a(s')ds'$$

Summing (integrating) over all such increments and adding on the transmitted beam $I_o e^{-t}$ gives

$$I = I_o e^{-\int_0^s k_a(s')ds'} + \int_0^s e^{-\int_{s'}^{s} k_a(s'')ds''} I_b(s')k_a(s')ds' \tag{7-5}$$

It should be noted that when radiators of different temperatures are present, for example, particles of various sizes, each size with a different temperature, one simply sums the k_a's to obtain the total k_a, and replaces I_b with an effective value given by

$$I_b = \frac{\sum_{i=1}^{n} k_{a,i} I_b(T_i)}{\sum_{i=1}^{n} k_{a,i}} \tag{7-6}$$

When isotropic scattering is a consideration, Eq. (5-8) applies instead of Eq. (5-6). Let the mean radiant intensity be S

$$S = \frac{1}{4\pi} \int_0^{4\pi} I d\Omega \tag{7-7}$$

Then Eq. (5-8) can be written

$$\frac{dI}{dt_e} = -I + (1-\omega_s) I_b + \omega_s S \tag{7-8}$$

where the optical depth for extinction is

$$t_e = \int_0^s (k_a + k_s) ds' \tag{7-9}$$

and ω_s is the "albedo for single scatter"

$$\omega_s = \frac{k_s}{k_a + k_s} \tag{7-10}$$

The quantity $1-\omega_s$ is sometimes given a name such as "particle emissivity". Comparison of Eq. (7-8) with Eq. (7-4) shows them to have the same form where the source function

$(1-\omega_s)I_b+\omega_s S$ replaces I_b. Thus Eq. (7-5) can be rewritten for nonzero k_s by replacing k_a with k_a+k_s and I_b with $(1-\omega_s)I_b+\omega_s S$

$$I = I_o e^{-\int_o^s (k_a+k_s)ds'} + \int_0^s e^{-\int_{s'}^s k_a(s'')ds''} \times [(1-\omega_s)I_b+\omega_s S](k_a+k_s)ds' \tag{7-11}$$

Equation (7-5) for $\omega_s = 0$ or Eq. (7-11) for $\omega_s \neq 0$ constitutes the formal mathematical solution to the equation of transfer for a nonhomogeneous and/or nonisothermal gas. The equation is seen to be directly useful when the source function is known as a function of position. For example, when one knows the temperature and soot-concentration distributions in a burner flame, one can find the radiant intensity emerging from the flame (neglecting scattering) by using Eq. (7-5) and carrying out the integrations numerically. Of course, if one wants the net flux, one integrates I over $\cos\theta\, d\Omega$ as in Eq. (6-1) or (1-8,9)

$$q_{net} = \int_0^{4\pi} I \cos\theta\, d\Omega = \int_0^{2\pi} (I^+-I^-)\cos\theta\, d\Omega \tag{7-12}$$

and, if one wants the volumetric heat source, one integrates I over $d\Omega$ and subtracts the volume emission given by Eq. (6-49)

$$\dot{Q}_{V,\text{net absorbed}} = k_a \int_0^{4\pi} I d\Omega - 4\pi k_a I_b = 4\pi k_a (S-I_b) \tag{7-13}$$

Often one wants not the intensity at the end of a path but the intensity incident at s = 0. The distance through the gas from s' to s = 0 is then s', and the transmitted intensity is $I_L e^{-t}$. Hence Eq. (7-5) becomes

$$I^-(0) = I_L e^{-\int_0^L k_a(s')ds'} + \int_0^L e^{-\int_0^{s'} k_a(s'')ds''} I_b(s')k_a(s')ds' \tag{7-14}$$

Also, Eq. (7-14) is often in a more convenient form for use when integration by parts is performed. Let $u = I_b(s')$ and

$$v = 1-e^{-\int_0^{s'} k_a(s'')ds''}, \quad dv = e^{-\int_0^{s'} k_a(s'')ds''} k_a ds'$$

Then

$$\int_0^L udv = uv\Big|_0^L - \int_0^L vdu = u_L v_L - \int_0^L vdu$$

Hence

$$I^-(0) = I_b(0) + [I_L - I_{bL}]e^{-\int_0^L k_a ds'}$$
$$+ \int_0^L e^{-\int_0^{s'} k_a ds''} \frac{dI_b}{ds'} ds' \tag{7-15}$$

If the backward intensity I^- is desired at a distance s along the path, one simply replaces each zero in Eq. (7-14) or Eq. (7-15) by s.

7.B Geometrical Considerations. The equation of transfer and its solution are simple. What complicates the subject of radiation heat transfer are geometrical complexities and spectral variations. We defer considering spectral variations and consider some geometrical effects. One may classify a problem as one-dimensional when the source function varies with one dimension only and multidimentional when it varies in more than one dimension. In the former case, four special geometries stand out, (1) the slab with variations in the z-direction only, (2) the sphere with variations in r only, (3) the cylinder with variations in r only, and (4) the cone with absorption coefficient varying with 1/z, where z is the distance from the apex, times a function of angle from the axis, and temperature varies with angle only. The slab may be applied to layers which are thin compared to any radius of curvature; the sphere to some chemical reactors, fireballs, and stellar interiors; the cylinder to pipes, combustion chambers, and cylindrical plumes; and the cone, as an idealization, to conical plumes and jets.

Even when a problem is one-dimensional in source function variation, the radiative transfer is three-dimensional in that rays in all directions contribute to the net flux. The slab and sphere possess azimuthal symmetry so that integration over solid angle reduces to integration over polar angle θ only, i.e., $d\Omega = 2\pi \sin\theta \, d\theta$. The cylinder and cone require integration over both base angle γ and axial angle β (recall Fig. 1-3).

Turning attention to the slab, one sees in Fig. 7-1 that the length of a ray varies from one slab thickness to infinity as angle θ varies from zero to $\pi/2$. Rays of all possible lengths between these limits contribute to the heat transfer. One can obtain satisfactory numerical answers by choosing a few discrete directions, attributing to each a finite solid angle, and replacing the integrals over solid angle with weighted summations. Such approximations are referred to as "discrete ordinates" or "multiflux" approximations. However, exact integral expressions can be employed for the simple geometries.

For the slab one simply recognizes that slant path length s is related to normal distance z by $s = z/\cos\theta$ as shown in Fig. 7-1. For the cylinder the trigonometric law of cosines is used to relate the change of radial distance r to the increment of path length ds, and, for a given ray and point of origin, s is double-valued in r. For a ray moving inwards, one integrates from the radius at the point of origin to r_{min} and then from r_{min} to r_w as illustrated in Fig. 7-1. Note that polar angle is appropriate for the sphere, but base-axial angle coordinates are more convenient for the cylinder and cone.

FIGURE 7-1

Relation of Path Length s to Position

The Slab

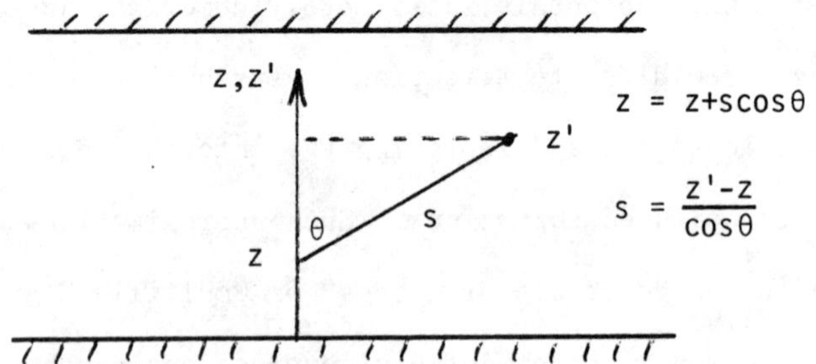

The Cylinder

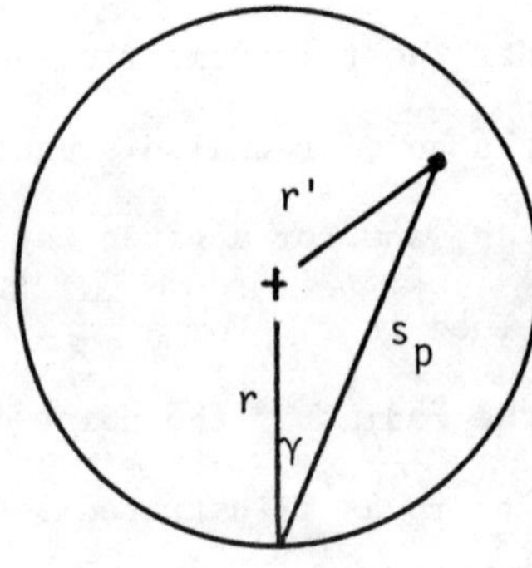

$$r'^2 = r^2+s_p^2-2rs_p\cos\gamma$$

$$s_p^2-2r\cos\gamma s_p+r^2-r'^2 = 0$$

$$s^2 = s_p^2+(z'-z)^2$$

(The quantity s_p is the projection of slant path s into the base plane.)

7.C The Slab Geometry. Introducing Eqs. (7-5) and (7-14) for location s into Eq. (7-12) gives rise to such terms as

$$\int_{\theta=0}^{\pi/2} e^{-k_a z/\cos\theta} \cos\theta \sin d\theta$$

One customarily denotes cos θ by μ or $1/\cos\theta$ by u. Hence a special function,

$$E_n(t) \equiv \int_0^1 e^{-t/\mu} \mu^{n-2} d\mu = \int_1^\infty e^{-ut} \frac{du}{u^n} \qquad (7\text{-}16)$$

is defined where in the example given $t = k_a z$. The exponential integral has properties from the definition as follows:

$$\frac{dE_n(t)}{dt} = E_n'(t) = -E_{n-1}(t) \;, \quad E_1'(t) = -e^{-t}/t \qquad (7\text{-}17)$$

$$\int E_n(t)dt = -E_{n+1}(t) \qquad (7\text{-}18)$$

$$E_n(0) = \frac{1}{n-1} \;, \quad E_1(t) = \ell n \frac{1}{t} - \gamma \quad \text{as} \quad t \to 0 \qquad (7\text{-}19)$$

$$\gamma = 0.5772156...$$

$$E_n(t) = \frac{1}{t} e^{-t} \quad \text{as} \quad t \to \infty \quad (t >> n) \qquad (7\text{-}20)$$

A short table of exponential integrals is given in Table 7-1.

Putting Eq. (7-5) for I^+ $(0 \leq \theta < \pi/2)$ and its equivalent for I^- $(\pi/2 < \theta \leq \pi)$ into Eq. (7-12) and identifying the exponential integrals gives, for diffuse boundaries,

$$q(t) = q_o^+ 2E_3(t) - q_L^+ 2E_3(t_L - t) + \int_0^t E_2(t-t')B(t')2dt' - \int_t^{t_L} E_2(t'-t)B(t')2dt' \qquad (7\text{-}21)$$

TABLE 7-1

VALUES OF THE FIRST FIVE EXPONENTIAL INTEGRALS

t	$E_1(t)$	$E_2(t)$	$E_3(t)$	$E_4(t)$	$E_5(t)$
0	∞	1.0000	0.5000	0.3333	0.2500
0.01	4.0379	0.9497	0.4903	0.3284	0.2467
0.02	3.3547	0.9131	0.4810	0.3235	0.2434
0.03	2.9591	0.8817	0.4720	0.3188	0.2402
0.04	2.6813	0.8535	0.4633	0.3141	0.2371
0.05	2.4679	.8278	.4549	.3095	0.2339
0.06	2.2953	.8040	.4468	.3050	.2309
0.08	2.0269	.7610	.4311	.2962	0.2249
0.10	1.8229	.7225	.4163	.2877	0.2190
0.20	1.2227	.5742	.3519	.2494	0.1922
0.30	0.9057	.4691	.3000	.2169	0.1689
0.40	0.7024	.3894	.2573	.1891	0.1487
0.50	0.5598	.3266	.2216	.1652	0.1310
0.60	0.4544	.2762	.1916	.1446	0.1155
0.80	0.3106	.2009	.1443	.1113	0.0901
1.00	0.2194	.1485	.1097	.0861	0.0705
2.00	0.0489	.0375	.0301	.0250	0.0213
3.00	0.0130	.0106	.0089	.0077	0.0067

Here q_o^+ is the radiosity of wall at z=0 (t=0 in terms of optical depth) and q_L^+ is the radiosity (directed in the negative z-direction) of the wall at z=L ($t=t_L$) and B is πI_b. Inspection of the equation and consideration of a differential slab of thickness dt' shows that $2E_3(t)$ is the hemispherical transmissivity for *diffuse* radiation through a slab of optical thickness t, and $1-2E_3(t)$ is the hemispherical absorptivity or (if the gas is isothermal) emissivity. The quantity $E_2(t-t')$ is the transmissivity for a *nondiffuse* source where $I = I_b k_a dz/\cos\theta$ from optical depth t' to optical depth t. The quantity 2dt' is the hemispherical emissivity of a slab of differential optical thickness $k_a dz'$. Recall that the mean beam length for an optically thin slab of thickness dz' is 2dz'; or, equivalently, the volume emission is $4k_a dz'B$, of which half goes forward and half backwards.

Just as Eq. (7-5) or (7-14) is often put in a more convenient form by integration by parts, so is Eq. (7-21). Let $v = 1-2E_3(t-t')$ and $u = B(t')$ in the first integral and $v' = 1-2E_3(t'-t)$ and $u' = B(t')$ in the second one. Then integration by parts gives

$$q(t) = [q_o^+ - B(0)]2E_3(t) - [q_L^+ - B(t_L)]2E_3(t_L - t) + \int_0^{t_L} 2E_3(|t-t'|)\left(-\frac{dB}{dt'}\right)dt' \qquad (7\text{-}22)$$

Note that, if B(t') is discontinuous, the integral in Eq. (7-22) is taken in the Stieltjes sense.

For example, consider a gas in a black flat plate duct of total thickness L with uniform absorption coefficient k_a whose temperature varies in such a way that the gas black body radiosity B rises linearly from the value B_w at each wall over boundary layer thickness δ_{BL} to a constant value of B_g throughout the remainder of the duct. Find the heat flux at the wall. With black walls and no temperature jump at the wall $q_o^+ - B(0) = 0$ and $q_L^+ - B(t_L) = 0$. Thus Eq. (7-22) at the wall becomes

$$q(0) = \int_0^{t_{BL}} 2E_3(t')[-\frac{B_g - B_w}{t_{BL}}]dt' + \int_{t_{BL}}^{t_L - t_{BL}} 2E_3(t')(0)dt'$$

$$+ \int_{t_L - t_{BL}}^{t_L} 2E_3(t')[+\frac{B_g - B_w}{t_{BL}}]dt'$$

$$-q(0) = \varepsilon_{channel}(B_g - B_w) \tag{7-23a}$$

$$\varepsilon_{channel} = \frac{1}{t_{BL}} \{\frac{2}{3}[1 - 3E_4(t_{BL})] - 2[E_4(t_L - t_{BL}) - E_4(t_L)]\} \tag{7-23b}$$

where $\varepsilon_{channel}$ is called the effective channel emissivity. One may easily insert some numbers; for example, if T_g = 1000 K, T_w = 500 K, L = 10 cm, δ_{BL} = 1 cm, and k_a = 100 m^{-1} (independent of λ over the spectrum of concern as defined by Table 1-1), the net radiant flux at the wall is

$$-q(0) = (5.67\text{x}10^{-8})(1000^4-500^4)\,\frac{1}{100(0.01)}$$

$$\times\ \{\tfrac{2}{3}\,[1-3E_4(1)]-2[E_4(9)-E_4(10)]\}$$

$$-q(0) = (5.316\text{x}10^4)(0.4945) = 2.628\text{x}10^4\ \text{W/m}^2$$

The example was chosen not just to illustrate Eq. (7-22) but also to give insight into a sometimes important phenomenon, self-absorption. In the numerical example the core of the gas between $t_{BL} = 1$ and $t_L - t_{BL} = 9$ was essentially opaque. Thus the irradiation q^- at the edge of the boundary layer was the full $B_g = \sigma T_g^{\ 4}$, and the radiosity at the wall $\sigma T_w^{\ 4}$ was $0.5^4 = 0.0625$ of the gas. Yet the net flux at the wall was only 0.4945 of the difference between $\sigma T_g^{\ 4}$ and $\sigma T_w^{\ 4}$, not $1-0.0625 = 0.9375$. In the boundary layer the irradiation upon the wall was reduced by absorption exceeding emission. For a fixed value of δ_{BL}/L one may increase $t_L = k_a L$ from zero to infinity. If δ_{BL}/L is zero one finds that the channel emissivity rises as $1-2E_3(t_L)$, linearly at first as $2t_L$ (since the geometric mean beam length is 2) and then more slowly to a maximum value of 1. When δ_{BL}/L is nonzero, one finds from Eq. (7-23b) that the channel emissivity rises linearly at first, as $(2-\delta_{BL}/L)t_L$, then more slowly until a maximum is reached, and then returns to zero, as $2/3(\delta_{BL}/L)t_L$, as t_L goes to infinity. Qualitatively, the same effect occurs in sooty flames from petroleum fires and in combustion chambers, that is, as the size of the flame increases, the radiant flux from it at first rises but then falls as the cold boundary layer becomes absorbing.

7.D <u>Differential Formulations</u>. When the source function is known, as it is in a nonscattering medium with known temperature profile, one simply integrates the equation of transfer along a path, to obtain I and integrates I over angles θ and ϕ (if required) or γ and β to find the heat flux. The integrations can be done numerically if necessary or with the aid of special functions such as the exponential integrals if they apply. When the source function is unknown, as for the case of appreciable scattering, or when radiative heating or cooling affects the temperature given by the general energy equation, one may be able to execute an iterative solution based upon estimating the source function, using the solution to the equation of transfer to find I, and improving the estimate of the source function by integrating I over 4π steradians and repeating. Such a procedure is convergent for albedo less than one in a medium of known temperature profile. Differential formulation may be a more convenient alternative. Some of the various differential methods are briefly described here as they would be applied to the classic engineering problem of predicting the radiant heat transfer through a layer of porous or fibrous insulation.

The situation considered is a layer of insulation L thick in radiative equilibrium. The material is imagined to be an isotropic gray emitter-absorber-scatterer with absorption coefficient $k = k_a + k_s$. Under conditions of gray radiative equilibrium ($dq_R/dz = 0$) there is no need to distinguish between isotropic scattering and absorption-reradiation. The bounding walls are taken as black. As for the adiabatic-walled passage, provision for nonblack surface emissivities can be made by adding surface resistances $(1-\varepsilon_1)/\varepsilon_1$ and $(1-\varepsilon_2)/\varepsilon_2$ to the insulation resistance $1/\mathscr{F}_{1-2,b}$ as shown in Eq. (3-59). Optical depth $t = kz$ is measured from one wall and varies from 0 to $t_L = kL$.

For comparison we can first derive the solution for the optically thin medium $t_L \ll 1$. For the optically thin case the medium is isothermal, and the network solution for a source, a sink, and an adiabatic gas applies with $\tau_g = e^{-2t_L} \doteq 1-2t_L$ and $\varepsilon_g = 1-\tau_g \doteq 2t_L$ where the factor of two appears because the mean beam length is 2L. The result is

$$\mathscr{F}_{1-2} = \tau_g + \varepsilon_g/2 = 1-t_L \quad \text{as} \quad t_L \rightarrow 0 \qquad (7\text{-}24)$$

The network solution clearly breaks down as t_L becomes large, because the assumption of an isothermal medium is then invalid.

7.D(a) Radiation Diffusion. At a wavelength where k_a is large so that (1) $1/k_a$ is short compared to any radius of curvature, (2) $k_a z_w$ is large where z_w is the distance to any wall, and (3) the quantity dB/dz is constant over a few multiples of $1/k_a$, where z is length measured in the direction of the temperature gradient, and B is πI_b of the medium at the wavelength of concern; Eq. (7-22) shows that the spectral radiant heat flux reduces to a simple expression in the form of Fourier's Law of heat conduction. The terms $2E_3(t)$ and $2E_3(t_L-t)$ go to zero because the optical depths t and t_L-t to the walls are large. The term $dB/dt' = (1/k_a)dB/dz'$ is regarded as constant at $(1/k_a)dB/dz$ over the range in which $2E_3(|t-t'|)$ drops to zero, and the integration with respect to t' is transformed to t-t' and t'-t. Since both t and t_L-t are large, the limits on the integrals are taken as effectively ∞. Then

$$q_\lambda = -\frac{1}{k_a}\frac{dB_\lambda}{dz}\left[4\int_0^\infty E_3(t'')dt''\right] = -\frac{4}{3k_a}\frac{\partial B_\lambda}{\partial T}\frac{dT}{dz} \qquad (7\text{-}25)$$

The quantity $(4/3k_a)(\partial B_\lambda/\partial T)$ is a spectral radiation conductivity. If large k_a applies to the entire spectrum of interest defined by temperature T and the internal fraction in Table 1-1, then one can integrate spectrally to obtain

$$q = -K_R\frac{dT}{dz} \qquad (7\text{-}26)$$

where

$$K_R = \int_0^\infty \frac{4}{3k_a}\frac{\partial B_\lambda}{\partial T}\,d\lambda = \frac{16}{3}\sigma T^3\int_0^1 \frac{1}{k_a}\,df_i(\lambda T) \qquad (7\text{-}27)$$

The quantity K_R, which is seen to be $\frac{16}{3}\sigma T^3$ times the internal-fraction-weighted average of $1/k_a$, is the Rosseland radiation conductivity.

The above relation was derived for a nonscattering medium. However, for a gray medium, isotropic scattering, and radiative equilibrium it is easy to see that the albedo of single scatter is of no concern (just as the emissivity of a perfectly diffuse gray refractory wall is of no concern), and k_a is replaced with the extinction coefficient k_a+k_s. For nongray anisotropic scattering the situation is not so simple, but an approximation is to use a $(1-\cos\theta)$-weighted average of the bidirectional scattering coefficient in place of k_s.

For the slab of insulation under consideration, one has merely

$$\frac{d}{dz}(q_R) = 0 \ , \quad q_R = -\frac{16}{3}\frac{1}{k}\sigma T^3 \frac{dT}{dz}$$

Recognizing that $4\sigma T^3 dT = dB$ permits writing

$$\frac{d}{dz}\left(\frac{4}{3k}\frac{dB}{dz}\right) = 0 \ , \quad \frac{d^2B}{dz^2} = 0$$

subject to $B = B_1$ at $z = 0$ and $B = B_2$ at $z = L$, if an "extrapolation length" is not applied. Obtaining the general solution for B, using the boundary conditions to solve for the constants, and finding $q_R = -(4/3k)dB/dz = \mathscr{F}_{1-2}(B_1-B_2)$ gives

$$\mathscr{F}_{1-2} = \frac{4}{3t_L} \quad \text{as} \quad t_L \to \infty \tag{7-28a}$$

The solution clearly breaks down as t_L goes to zero.

To improve the accuracy of the radiation diffusion approximation, the notion of an extrapolation length is sometimes applied. Writing Eq. (7-22) at the wall z = 0 and letting k become large gives $q_R = -(2/3k)\partial B/\partial z$ vs $-(4/3k)\partial B/\partial z$ some distance from the wall. Extrapolating the far-from-wall slope to the wall value suggests imposing B_W at distance 2/3k behind the wall. In the present example L is replaced by L + 4/3k giving

$$\mathscr{F}_{1-2} \doteq \frac{1}{1 + \frac{3}{4} t_L}, \quad t_L >> 1 \tag{7-28b}$$

This result is a good engineering approximation. However, it is not clear how to use the extrapolation length concept in a combined radiation and conduction-convection problem.

7.D(b) Multiflux Methods. The idea is simply to choose a few discrete directions (sometimes called "discrete ordinates") and use the equation of transfer to give intensities in those directions and then form approximate expressions for S and q_R by summing them ("finite quadrature"). For the slab of insulation under consideration the lowest level of this approximation is the Schuster two flux model, which is equivalent to using a so-called "substitute kernel", i.e., approximating $E_2(t)$ with e^{-at}. The choice of directions is arbitrary. At the two-flux level, one might choose the directions back and forth along the geometric mean beam length (which would give correct results as $t_L \to 0$) or a somewhat shorter slant path. At the two-flux level we might number the two intensities as I_1 and I_2 and the fluxes as q_1 and q_2. The net flux would be approximated with

$$q_R \doteq \pi I_1 - \pi I_2 = q_1 - q_2$$

and the source function with

$$S \doteq \frac{1}{2}(I_1+I_2) = \frac{1}{2\pi}(q_1+q_2)$$

At radiative equilibrium $I_b = S$, and the equation of transfer for I_1 and I_2 is

$$\frac{dI_1}{dt_1} = -I_1+S \qquad \frac{dq_1}{dt_1} = -q_1 + \frac{1}{2}(q_1+q_2)$$

$$\frac{dI_2}{dt_2} = -I_2+S \qquad \frac{dq_2}{dt_2} = -q_2 + \frac{1}{2}(q_1+q_2)$$

We let $t_1 = akz = at$, $t_2 = -akz = -t_1$, and choose $a = 2$, thus aligning I_1 and I_2 along the geometric mean beam length. The two-flux equations become

$$\frac{dq_1}{dt} = q_2-q_1 = q_R \text{ , } \frac{dq_2}{dt} = q_2-q_1 = q_R$$

subject to $q_1 = B_1$ at $t = 0$ and $q_2 = B_2$ at $t = t_L$. Solving the two equations simultaneously and imposing the boundary conditions to solve for unknown constants gives q_1 and q_2.

Then the net flux is found. The result for $\mathscr{F}_{1-2}$ is

$$\mathscr{F}_{1-2} = \frac{1}{1+t_L} \tag{7-29}$$

The result is seen to be correct as $t_L \to 0$, because the geometric mean beam length was chosen, but comparison to Eq. (7-28a) shows it underpredicts by 25% as $t_L \to \infty$.

If, say, six fluxes are used, the accuracy of the multi-flux method becomes quite acceptable for engineering purposes. With the selection of 2n fluxes one must choose 2n directions and attribute to each direction a weighting coefficient a_i for the net flux and b_i for the source function (i = 1,n).

$$q = \sum_{i=1}^{2n} a_i q_i \ , \quad S = \sum_{i=1}^{2n} b_i q_i \tag{7-30}$$

A primitive selection is to use equally spaced values of μ^2 ($\mu = \cos\theta$) over the interval 0 to +1, i.e.,

$$\mu_i = [1-(i-1/2)/n]^{\frac{1}{2}} \ , \ 1 \leq i \leq n \ ;$$

$$\mu_i = -[(i-1/2)-1]^{\frac{1}{2}} \ , \quad n+1 \leq i \leq 2n \tag{7-31}$$

Then, because $\cos\theta \, d\Omega/\pi$ is $d\mu^2$ and $d\Omega/4\pi$ is $d\mu$

$$a_i = 1/n \ , \quad b_i = [1-(i-1)/n]^{\frac{1}{2}} - [1-i/n]^{\frac{1}{2}} \ , \quad 1 \leq i \leq n$$

$$a_i = -1/n \ , \quad b_i = [i/n-1]^{\frac{1}{2}} - [(i-1)/n-1]^{\frac{1}{2}} \ , \quad n+1 \leq i \leq 2n \tag{7-32}$$

Other selection schemes have been examined, e.g. [8].

7.D(c) Embedding. The embedding concept is used to formulate differential equations for the bidirectional reflectivity and transmissivity in conjunction with the use of discrete ordinates. Consider the problem of finding the bidirectional reflectivity and transmissivity of a layer of snow. The problem is related to our layer of insulation because in the limit as albedo for single scatter ω_s goes to unity, the scattering layer becomes the same as the scattering-reradiating layer in radiative equilibrium. The layer then absorbs nothing, and the hemispherical transmissivity becomes one minus the hemispherical reflectivity. The hemispherical transmissivity is thus identical to $\mathscr{F}_{1-2,b}$ of a layer in radiative equilibrium.

Formulating by embedding is accomplished by imagining that the bidirectional properties of a slab of optical depth t are known and then asking what change would occur upon adding an infinitesimal layer dt. The layer can be imagined as added to the top (the irradiated side) or the bottom (the dark side). The procedure is referred to as embedding from above or embedding from below accordingly. Figure 7-2 shows the situation for calculating the reflectivity by embedding from above.

Because dt is infinitesimal, only first order terms are retained. The figure shows all the first order rays and the

FIGURE 7-2

Embedding from Above, Reflectivity

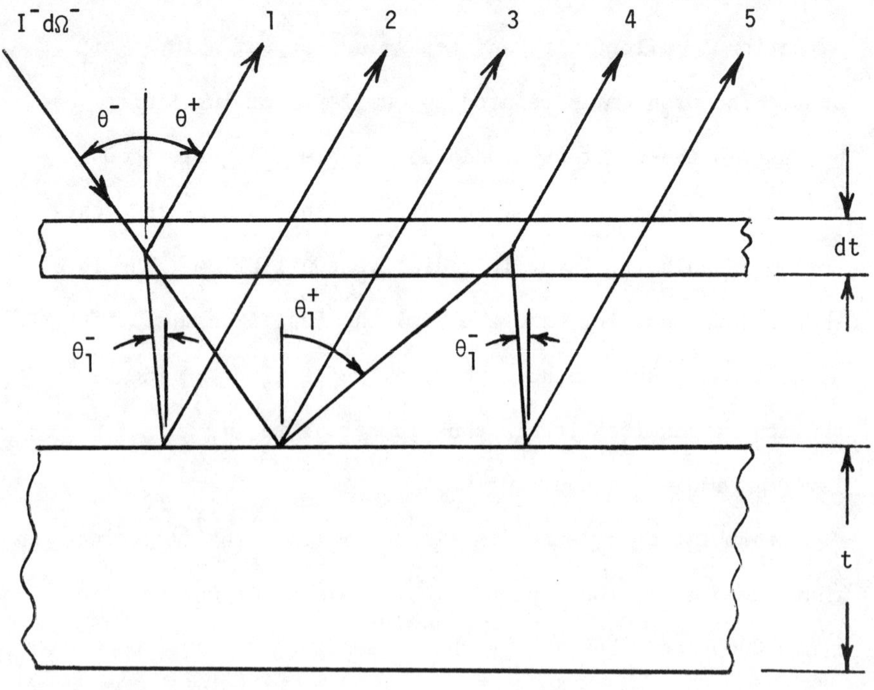

zeroth order one. The slab is irradiated by intensity I^- in solid angle $d\Omega^-$ about angle θ^-. We want to know the intensity I^+ in direction θ^+. The five rays shown are all parallel in the emergent direction of interest θ^+. The first was scattered directly into the θ^+ direction. The slant path in that direction is $dt/\cos\theta^+$ and the source strength in the slab is $\omega_s I^- d\Omega^-/4\pi$. Hence

$$dI_1^+ = \frac{\omega_s}{4\pi} I^- d\Omega^- \frac{dt}{\cos\theta^+} \tag{7-33}$$

The second was scattered downwards into some arbitrary direction θ_1^- and then bidirectionally reflected into the desired angle θ^+. The second order attenuation by the thin slab dt is neglected. From the definition of the bidirectional reflectivity Eq. (2-14)

$$dI_2^+ = \frac{1}{\pi}\rho(\theta^+,\theta_1^-)dI_1^-\cos\theta_1^- d\Omega_1^-$$

The downward scattered intensity dI_1^- is of the form of Eq. (7-33) with the angle θ_1^- replacing θ^+. We integrate over all possible θ_1^- directions. Accordingly

$$dI_2^+ = [\frac{1}{\pi}\int_{2\pi}\rho(\theta^+,\theta_1^-)d\Omega_1^-]\frac{\omega_s}{4\pi} I^- d\Omega^- dt \tag{7-34}$$

where

$$d\Omega_1^- = 2\pi\sin\theta_1^- d\theta_1^- = 2\pi d(\cos\theta_1^-)$$

The third ray was transmitted through the added layer dt, bidirectionally reflected into the desired angle θ^+ and transmitted again through dt. Upon each transmission an attenuation occurs.

$$I_3^+ = (1 - \frac{dt}{\cos \theta^+}) \frac{1}{\pi} \rho(\theta^+, \theta^-)(1 - \frac{dt}{\cos \theta^-}) I^- d\Omega^-$$

Upon neglect of the dt^2 term

$$I_3^+ = \frac{1}{\pi} \rho(\theta^+, \theta^-) I^- d\Omega^- [1 - (\frac{1}{\cos \theta^+} + \frac{1}{\cos \theta^-}) dt] \quad (7\text{-}35)$$

This contribution is seen to contain the zeroth order term and a first order one.

Rays 4 and 5 arise from bidirectional reflection of the transmitted beam and subsequent scattering. For ray 4 the scattering is into the direction of interest θ^+, and for ray 5 the scattering is downwards at angle θ_1^- followed by bidirectional reflection into the desired direction θ^+. Since first order scattering occurs, the higher order attenuations are neglected. Integration over possible scattering angles is made.

$$dI_4^+ = \frac{1}{\pi} \int_{2\pi} \rho(\theta_1^+, \theta^-) d\Omega_1^+ \; I^- \cos \theta^- d\Omega^- \frac{\omega_s dt}{4\pi \cos \theta^+} \quad (7\text{-}36)$$

$$dI_5^+ = \frac{1}{\pi} \int_{2\pi} \rho(\theta_1^+, \theta^-) d\Omega_1^+ \; I^- \cos \theta^- d\Omega^- (\frac{\omega_s dt}{4\pi})$$

$$\times \frac{1}{\pi} \int_{2\pi} \rho(\theta^+, \theta_1^-) d\Omega_1^- \quad (7\text{-}37)$$

Summing the 5 terms and dividing by $(I^- \cos \theta^- d\Omega^- / \pi)$ gives $\rho(\theta^+, \theta^-)$ for thickness t+dt. Denoting $\cos \theta^+$ as u, $\cos \theta^-$

as v, cos θ_1^+ as u' and cos θ_1^- as v'; subtracting $\rho(u,v,t)$; dividing by dt; and taking the limit as dt goes to zero gives

$$\frac{\partial\rho(u,v,t)}{\partial t} = \omega_s \left\{\frac{1}{2u} + \int_0^1 \rho(u,v',t)dv'\right\} \left\{\frac{1}{2v} + \int_0^1 \rho(u',v,t)du'\right\} - (\frac{1}{u} + \frac{1}{v})\rho(u,v,t) \qquad (7\text{-}38)$$

Choosing n discrete directions, say those given by Eq. (7-31), and understanding a_i and b_i to be as given by Eq. (7-32) gives n^2 simultaneous ordinary differential equations

$$\frac{\partial\rho_{i,j}(t)}{\partial t} = \omega_s \left\{\frac{1}{2u_i} + \sum_{k=1}^{n} b_k\rho_{i,k}(t)\right\} \left\{\frac{1}{2v_j} + \sum_{k=1}^{n} b_k\rho_{k,j}(t)\right\} - (\frac{1}{u_i} + \frac{1}{v_j})\rho_{i,j}(t) \qquad (7\text{-}39)$$

These are solved with a forward-marching Runge-Kutta integration commencing with $\rho_{i,j}(0) = 0$ for a black boundary. Once $\rho_{i,j}(t_L)$ is found, the hemispherical reflectance is

$$\rho_H(t_L) = \sum_{i=1}^{n} \sum_{j=1}^{n} a_i a_j \rho_{i,j}(t_L) \qquad (7\text{-}40)$$

and for $\omega_s = 1$

$$\mathscr{F}_{1-2,b} = 1-\rho_H(t_L) \qquad (7\text{-}41)$$

7.D(d) Other Differential Methods. Various other approximate methods have been used to analyze radiative transfer with scattering. In the classical Milne Eddington approximation the equation of transfer, Eq. (7-8) with

$dt_e = kds = kdz/\mu = dt/\mu$, is integrated with respect to $d\Omega$ over 4π steradians and then with respect to $\cos\theta \, d\Omega$ over the same interval. The two integrations are the zeroth and first moment, respectively. Hence the method is a "moment method". The approximation is made that

$$K = \int_{-1}^{+1} I\mu^2 d\mu \doteq \frac{2}{3}\left(\frac{1}{2}\int_{-1}^{+1} Id\mu\right) = \frac{2}{3} S \qquad (7\text{-}42)$$

Thus two simultaneous equations are obtained in net flux q_R and source function S

$$\frac{dq_R}{dt} = 4\pi(1-\omega_s)(I_b - S) \qquad (7\text{-}43)$$

$$\frac{dS}{dt} \doteq -\frac{3}{4\pi} q_R \qquad (7\text{-}44)$$

For the insulation problem proposed, radiative equilibrium applies, and $I_b = S$ so Eq. (7-43) becomes

$$\frac{dq_R}{dt} = 0 \qquad (7\text{-}43a)$$

The two equations can be solved at once to show q_R is a constant and S a linear function in t,

$$S = S_o - \frac{3}{4\pi} q_R t \qquad (7\text{-}45)$$

The two unknown constants q_R and S_o must be found from boundary conditions. At this point, as is characteristic with approximate differential methods, the problem becomes somewhat complex. One returns to the equation of transfer and solves for I_L^+ and I_o^-, Eqs. (7-11) and (7-14) respectively; and, knowing $I_o^+ = B_1/\pi$ and $I_L^- = B_2/\pi$, finds S at $t = 0$ and $t = t_L$ by integrating with respect to μ; and equates them to Eq. (7-45), thus obtaining two simultaneous equations in the two unknowns. Other moment methods have been explored [10].

7.E <u>Spectral Considerations, Scaling Approximations</u>. We turn our attention now away from geometrical (directional and spatial) considerations to consideration of spectral variations. We have seen that the equation of transfer is readily solved, e.g. Eq. (7-5), on a true spectral basis. The solution is useful as it stands when one is dealing with a continuum source such as soot or particulates. But when the source is a molecular gas, the spectral absorption coefficient k_a may vary by several orders of magnitude over the spectral region between lines and over a band. In such a case, Eq. (7-5) or (7-14) is of no practical use. Recall that one often has no knowledge of k_a but only of narrow band model parameters S/d (line-intensity-to-spacing ratio) and β (line-width-to-spacing ratio parameter) and in some cases no knowledge of those parameters but only of wide band parameters α_k (integrated band intensity), β_k (line-width-to-spacing ratio parameter), and ω_k (band width parameter). Accordingly one resorts to what is called "narrow band scaling" or "wide band scaling" to arrive at working relations.

7.E(a) Narrow Band Scaling: The Curtis-Godson Approximation. Consider the spectrally smoothed transmissivity of a path L of gas whose temperature, composition, and/or pressure varies along L. From Eqs. (7-2), (7-3), and (5-18)

$$\bar{\tau}_g = \frac{1}{d}\int_{\nu-d/2}^{\nu+d/2} e^{-\int_0^X k_a dX} d\nu \qquad (7\text{-}46)$$

where k_a is given by a sum of Lorentz lines according to Eq. (5-15) and X is the path-length-absorber gas density product. For the homogeneous gas where k_a does not vary along the path, the smoothed transmissivity $\bar{\tau}_g$ was given by an equation such as Eq. (5-19) as a function of S/d and γ/d or $\eta = \pi\gamma/d$, where S/d was the spectral average of k_a over the d interval. It is clear that for the optically thin nonhomogeneous gas, where $k_a L << 1$ everywhere in the spectral range ν-d/z to ν+d/2, that

$$\bar{\tau}_g = \frac{1}{d}\int_{\nu-d/2}^{\nu+d/2} [1-\int_o^X k_a dX]d\nu = 1 - \int_0^X (S/d)dX = 1-\tilde{\xi}$$

while the corresponding relation for the homogeneous gas is

$$\bar{\tau}_{g,homo} = 1 - (S/d)X$$

Hence, for the optically thin nonhomogeneous gas

$$\bar{\tau}_g = \bar{\tau}_{g,homo}(\xi=\tilde{\xi})$$

where ξ is (S/d)X for the homogeneous gas and

$$\tilde{\xi} = \int_0^X (S/d)dX = (\tilde{S}/d)X \qquad (7\text{-}47)$$

When the lines are strong, that is, optically thick near $\nu = \nu_i$ in Eq. (5-15), the transmissivity is low there. Appreciable transmissivity is contributed only when $(\nu-\nu_{o,i})^2$ greatly exceeds γ^2. Under this condition comparison of Eq. (7-46) for the nonhomogeneous gas with that for the homogeneous gas suggests that

$$\bar{\tau}_g = \bar{\tau}_{g,homo}(\zeta=\tilde{\zeta}) \tag{7-48}$$

$$\tilde{\zeta} = \int_0^X (S/d)(\pi\gamma/d)dX = (\tilde{S}/d)(\tilde{\eta})X \tag{7-49}$$

The Curtis-Godson approximation is to assume that

$$\bar{\tau}_g \doteq \bar{\tau}_{g,homo}(\xi=\tilde{\xi},\zeta=\tilde{\zeta}) \tag{7-50}$$

In terms of Eq. (5-19) the Curtis-Godson approximation is

$$\bar{\tau}_g \doteq \exp \frac{-\tilde{\xi}}{[1+\tilde{\xi}^2/\tilde{\zeta}]^{1/2}} \tag{7-51}$$

To obtain a working formula for nonisothermal gas intensity one spectrally smooths Eq. (7-5) or (7-14). Comparison of the spectrally smoothed Eq. (7-14) with Eq. (7-46) shows that

$$\bar{I}_{sm}(0) = I_L\bar{\tau}_g + \int_0^{L'} [-\frac{d\bar{\tau}_g}{ds}] I_b(s)ds, \quad L' = L/\cos\theta \tag{7-52}$$

Here it has been assumed that I_L is spectrally slowly varying, which really restricts the development to low reflecting boundaries. Integration by parts (or spectrally smoothing Eq. (7-15)) gives

$$I^-_{sm}(0) = I_b(0) + [I_L - I_b(L)]\bar{\tau}_g(L) + \int_0^{L'} \bar{\tau}_g dI_b(s) \quad (7\text{-}53)$$

Now it is clear how to make a practical calculation having narrow band model data S/d and $\eta = \pi\gamma/d$ versus ν and T_g. One marches along a line of sight computing $\tilde{\xi}$ and $\tilde{\zeta}$, $\bar{\tau}_g(\tilde{\xi},\tilde{\zeta})$ and $\int \tau_g dI_b(s)$ as one goes. When the end of the path is reached, the needed values are substituted into Eq. (7-53).

7.E(b) Wide Band Scaling. Equation (7-53) can be put in the form

$$I^-_{sm}(0) = I_L - [I_b(L) - I_L][1-\bar{\tau}_g] - \int_0^{L'} [1-\bar{\tau}_g]dI_b(s) \quad (7\text{-}54)$$

Integration over all wavenumbers and introduction of the band approximation (recall Eqs. (5-26) and (6-15)) gives the total intensity incident upon the wall

$$I^-_T(0) = I_{L,T} + \Sigma_k[I_b(\nu_k,T_L) - I_L(\nu_k)]A_k(L) - \Sigma_k \int_0^{L'} A_k(s)dI_b(\nu_k,s) \quad (7\text{-}55)$$

where the subscript T denotes a total quantity ($I_{L,T} = \sigma T_L^4/\pi$) and where A_k is the band absorption for the nonisothermal gas. Use of Eq. (7-55) depends upon being able to calculate the band absorption A_k of a nonisothermal gas path.

A wide-band-scaling rule proposed by Edwards and Morizumi [13] is to find ξ_1, ξ_2, and ξ_3 and compute $\tilde{\alpha}$, $\tilde{\omega}$, and $\tilde{\eta}$ which are used in place of α, ω, and η in Eq. (5-58) to find band absorption A

$$\xi_{1,k} = \int_0^s \alpha_k \rho_a ds = \int_0^X \alpha_k dX = \tilde{\alpha}_k X \tag{7-56a}$$

$$\xi_{2,k} = \int_0^s \omega_k \alpha_k \rho_a ds = \int_0^{\xi_{1,k}} \omega_k d\xi_{1,k} = \tilde{\omega}_k \xi_{1,k} \tag{7-56b}$$

$$\xi_{3,k} = \int_0^s \eta_k \omega_k \alpha_k \rho_a ds = \int_0^{\xi_{2,k}} \eta_k d\xi_{2,k} = \tilde{\eta}_k \xi_{2,k} \tag{7-56c}$$

In a test comparison with actual nonisothermal gas emission data, the wide-band-spacing rule gave good engineering accuracy [13].

7.F Molecular Gas Radiation in the Slab. Consider now the problem of accounting for the combined effects of spectral and directional variations in a nonisothermal slab of molecular gas. One must integrate over both the spectrum and the forward and backward hemispheres to find the net radiant flux.

$$q_R = \int_{\nu=0}^{\infty} \int_{\mu=0}^{1} [I_\nu^+(\mu) - I_\nu^-(\mu)] 2\pi \, \mu d\mu \, d\nu \tag{7-57}$$

Thus far only one or the other of the two integrations has been considered, the hemispherical integration yielding Eq. (7-22) and the spectral integration (at the wall) yielding Eq. (7-55). Since the order of the integrations is interchangeable, one can find the desired relation for q_R by either integrating Eq. (7-22) spectrally or Eq. (7-55) (generalized to an arbitrary location) hemispherically. We choose the latter route because Eq. (7-55) already incorporates the idea of wide-band scaling, while Eq. (7-22) does not.

7.F(a) Notation. To generalize Eq. (7-55) to an arbitrary location y measured from one wall, we must consider paths from y' to y. Hence Eqs. (7-56a-c) are written

$$\xi_{1,k}(y,y') = \int_{y'}^{y} \alpha_k \rho_a dy'' = \xi_{1,k}(y) - \xi_{1,k}(y') \qquad (7\text{-}58a)$$

$$\xi_{2,k}(y,y') = \int_{y'}^{y} \omega_k \alpha_k \rho_a dy'' = \xi_{2,k}(y) - \xi_{2,k}(y') \qquad (7\text{-}58b)$$

$$\xi_{3,k}(y,y') = \int_{y'}^{y} \eta_k \omega_k \alpha_k \rho_a dy'' = \xi_{3,k}(y) - \xi_{3,k}(y') \qquad (7\text{-}58c)$$

The band absorption for a slant path inclined by angle θ ($\mu = \cos\theta$) between y' and y is then

$$A_{k,\mu}(y,y') = \tilde{\omega}_k A_k^*(\tilde{t}_k/\mu, \tilde{\eta}_k) \qquad (7\text{-}59)$$

where

$$\tilde{\omega}_k = \xi_{2,k}(y,y')/\xi_{1,k}(y,y') \qquad (7\text{-}60a)$$

$$\tilde{t}_k = \xi_{1,k}^2(y,y')/\xi_{2,k}(y,y') \qquad (7\text{-}60b)$$

$$\tilde{\eta}_k = \xi_{3,k}(y,y')/\xi_{2,k}(y,y') \qquad (7\text{-}60c)$$

Recall that the band absorptance A_k^* for the exponential-tailed band model is defined by Eqs. (5-19), (5-23), and (5-24) as follows:

$$A_k^*(t_k, \eta_k) = \int_0^\infty \left\{ 1 - \exp\left[\frac{-t_k e^{-\nu^*}}{\sqrt{1+(t_k/\eta_k)e^{-\nu^*}}} \right] \right\} d\nu^* \qquad (7\text{-}61)$$

where $\nu^* = (\nu-\nu_k)/\omega_k$. When η_k is large, as it is at high temperatures because of the behavior of Eqs. (5-47,48), Eq. (7-61) reduces to

$$A_k^*(t_k) = \ln t_k + E_1(t_k) + \gamma \ , \ \eta_k >> 1 \qquad (7\text{-}62)$$

Recall also the discussion below Eq. (5-58); when Eq. (7-62) is used the values of ω_k from Table 5-1 are to be increased by 20 per cent.

7.F(b) Forward and Backward Intensities. With y understood to go from 0 at one wall to L at the other, Eq. (7-55) is generalized to location y by integrating from y to L instead of from 0 to L. Subscript T for total is dropped to simplify notation

$$I_\mu^-(y) = I_L + \Sigma_k [I_{b,k,L} - I_{L,k}] A_{k,\mu}(L,y) - \Sigma_k \int_y^L A_{k,\mu}(y',y) \frac{dI_{b,k}}{dy'} dy' \qquad (7\text{-}63)$$

Provision for the physically unreal idea of discontinuities in $I_{b,k}$ with y' is dropped.

The forward intensity follows from Eq. (7-63) upon reversing the coordinate system from y to L-y.

$$I_\mu^+(y) = I_0 + \Sigma_k [I_{b,k,0} - I_{0,k}] A_{k,\mu}(y,0) + \Sigma_k \int_0^y A_{k,\mu}(y,y') \frac{dI_{b,k}}{dy'} dy' \qquad (7\text{-}64)$$

7.F(c) Slab Band Absorptance. Integration of $(I^+ - I^-)$ over μ converts the band absorption A_k to slab band absorption A_s. The corresponding dimensionless quantities are band absorptance A_k^* and slab band absorptance A_s^*.

$$A_s(t_k, \eta_k) = \int_0^1 A_k(t_k/\mu, \eta_k) 2\mu d\mu \tag{7-65a}$$

$$A_s^*(t_k, \eta_k) = \int_0^1 A_k^*(t_k/\mu, \eta_k) 2\mu d\mu \tag{7-65b}$$

Substitution of Eq. (7-62) into Eq. (7-65b) gives the result for large η_k

$$A_s^*(t_k) = \ln t_k + E_1(t_k) + \gamma + \frac{1}{2} - E_3(t_k) \qquad \eta_k \gg 1 \tag{7-66}$$

For diffuse walls, substitution of Eq. (7-64) minus (7-63) for

$$\int_0^\infty (I_\nu^+ - I_\nu^-) d\nu$$

into Eq. (7-57) then gives

$$\begin{aligned} q_R(y) = q_0^+ - q_L^+ &+ \Sigma_k [B_{k,0} - q_{k,0}^+] A_s(y,0) \\ &- \Sigma_k [B_{k,L} - q_{k,L}^+] A_s(L,y) \\ &+ \Sigma_k \int_0^y A_s(y,y') \frac{dB_k}{dy'} dy' \\ &+ \Sigma_k \int_y^L A_s(y',y) \frac{dB_k}{dy'} dy' \end{aligned} \tag{7-67}$$

where the notation $A_s(y,y')$ has the meaning

$$A_s(y,y') = \tilde{\omega}_k A_s^*(\tilde{t}_k,\tilde{\eta}_k) \tag{7-68}$$

with $\tilde{\omega}_k$, $\tilde{t}_k$, and $\tilde{\eta}_k$ given by Eqs. (7-60a-c).

7.F(d) <u>The Symmetrical Black-Walled Slab</u>. For black walls at identical temperatures only the last two terms of Eq. (7-67) remain. If in addition symmetry exists in gas temperature about the center $y = \delta = L/2$, i.e.,

$$B_k(y') = B_k(y'') \ , \quad y'' = L-y' \tag{7-69}$$

the integral from y to L is written from y to δ, δ to 2δ-y, and 2δ-y to L = 2δ. Consider the integral from δ to 2δ-y,

$$\int_{\delta}^{2\delta-y} A_s(y',y)\,\frac{dB_k}{dy'}\,dy' = -\int_y^{\delta} A_s(2\delta-y'',y)\,\frac{dB_k}{dy''}\,dy''$$

Similarly

$$\int_{2\delta-y}^{2\delta} A_s(y',y)\,\frac{dB_k}{dy'}\,dy' = -\int_0^{y} A_s(2\delta-y'',y)\,\frac{dB_k}{dy''}\,dy''$$

Thus Eq. (7-67) becomes

$$q_R(y) = -\Sigma_k\left\{\int_0^y [A_s(2\delta-y'',y)-A_s(y,y'')]\,\frac{dB_k}{dy''}\,dy'' + \int_y^{\delta} [A_s(2\delta-y'',y)-A_s(y'',y)]\,\frac{dB_k}{dy''}\,dy''\right\} \tag{7-70}$$

It remains to specify $A_s(2\delta-y'',y)$, $A_s(y,y'')$, and $A_s(y'',y)$ in terms of ξ integrals defined over the half channel. With reference to Eq. (7-58) it is clear that the integral from y to 2δ-y" is the difference of integrals from 0 to δ and

0 to y plus the difference of 0 to δ and 0 to y''. Hence

$$\xi_{i,k}(2\delta-y'',y) = 2\xi_{i,k}(\delta)-\xi_{i,k}(y)-\xi_{i,k}(y'') \qquad (7\text{-}71a)$$

$$\xi_{i,k}(y,y'') = \xi_{i,k}(y)-\xi_{i,k}(y'') \qquad (7\text{-}71b)$$

$$\xi_{i,k}(y'',y) = \xi_{i,k}(y'')-\xi_{i,k}(y) \qquad (7\text{-}71c)$$

With this clarification Eqs. (7-60) and (7-68) give the desired terms.

7.F(e) Channel Emissivity. Consider a black-walled channel containing a gas with a trapezoidal source profile, B rising linearly from B_w over distance δ_{BL} to a constant value B_g in the range $\delta_{BL} \leq y \leq L-\delta_{BL}$ and then decreasing symmetrically from B_g back to B_w. Assume that $(T_g-T_w)/T_w << 1$ so *all* of the $B(\nu_k,T(y))$ vary with y in the trapezoidal way proposed. Consistent with this assumption we consider α_k, β_k, and ω_k are independent of y, and we take η_k large. Accordingly

$$\tilde{\omega}_k = \omega_k \qquad (7\text{-}72a)$$

$$\tilde{t}_k(2\delta-y'',y) = t_{L,k}[1-y^*-y^{*''}] \qquad (7\text{-}72b)$$

$$\tilde{t}_k(y,y') = t_{L,k}[y^*-y^{*'}] \qquad (7\text{-}72c)$$

where

$$t_{L,k} = \frac{\alpha_k \rho_a L}{\omega_k}, \quad y^* = \frac{y}{L}, \quad y^{*'} = \frac{y'}{L} \qquad (7\text{-}72d\text{-}f)$$

The wall heat flux is given by Eq. (7-70) at y = 0, and with $dB_k/dy'' = (B_g-B_w)/\delta_{BL}$ in the 0 to δ_{BL} interval and zero elsewhere

$$q_R(0) = -\Sigma_k \frac{B_{g,k}-B_{w,k}}{\delta_{BL}} \int_0^{\delta_{BL}} [A_s(t_{L,k}(1-y^{*''})) - A_s(t_{L,k}y^{*''})]dy''$$

Changing variables of integration from y'' to y*'' to t_k'' gives

$$q_R(0) = -\Sigma_k \frac{\omega_k(B_{g,k}-B_{w,k})}{t_{BL,k}} \int_0^{t_{BL,k}} [A_s^*(t_{L,k}-t_k'')-A_s^*(t_k'')]dt_k''$$

The integrals require knowing

$$S^*(t) = \int_0^t A_s^*(t')dt' \tag{7-73}$$

From Eq. (7-66) for large η_k

$$S^*(t) = t[\ell nt+\gamma-\tfrac{1}{2}]+[1-E_2(t)]-[\tfrac{1}{3} - E_4(t)] \tag{7-74}$$

Hence

$$q_R(0) = -\Sigma_k\omega_k[B(\nu_k,T_g)-B(\nu_k,T_w)]C^*(t_{L,k},t_{BL,k}) \tag{7-75a}$$

where

$$C^*(t_L,t_{BL}) = \frac{1}{t_{BL}}\left\{S^*(t_L)-S^*(t_L-t_{BL})-S^*(t_{BL})\right\} \tag{7-75b}$$

Table 7-2 lists values of channel emissivity for various pairs of t_L and t_{BL}. For comparison there is shown Eq. (7-23) for spectral radiation at the band head (or for a gray band). For a fixed value of δ_{BL}/L, as one increases t_L from zero to infinity by increasing the density of absorber ρ_a, Eq. (7-75) indicates the heat flux rises monotonically to an asymptotic limit, while Eq. (7-23) rises to a maximum and then falls to

TABLE 7-2

CHANNEL EMISSIVITY AND WALL LAYER TRANSMISSIVITY
FOR AN EXPONENTIAL-TAILED GAS BAND
WITH OVERLAPPED LINES

Optical Depth at Band Head		Channel Emissivity		Wall Layer Transmissivity	
Across Channel	Across One Boundary Layer	Exponential Band	Spectral Value at Band Head	Exponential Band	Spectral Value at Band Head
t_L	t_{BL}	C^*	$\varepsilon_{channel}$	τ_{WL}	$\tau_{WL,\lambda}$
2.0	0.0	1.789	0.940	1.00	1.00
	0.1	1.672	0.848	0.98	0.95
	0.2	1.563	0.770	0.97	0.91
	0.4	1.362	0.643	0.95	0.86
	1.0	0.832	0.372	0.93	0.79
20.0	0.0	4.073	1.000	1.00	1.00
	0.1	3.977	0.913	.98	0.92
	0.2	3.891	0.839	.96	0.85
	0.4	3.736	0.721	.94	0.74
	1.0	3.366	0.494	.87	0.52
	2.0	2.924	0.308	.80	0.34
	4.0	2.335	0.167	.72	0.21
	10.0	1.320	0.067	.65	0.13
200.0	0.0	6.376	1.000	1.00	1.00
	0.1	6.282	0.913	.99	0.91
	0.2	6.198	0.839	.97	0.84
	0.4	6.048	0.721	.95	0.72
	1.0	5.692	0.494	.90	0.50
	2.0	5.273	0.308	.84	0.31
	4.0	4.735	0.167	.76	0.17
	10.0	3.904	0.067	.64	0.07
	20.0	3.218	0.033	.56	0.04
	40.0	2.485	0.017	.49	0.02
	100.0	1.380	0.007	.43	0.01

zero. The difference in total band behavior and spectral behavior arises from the fact that the spectral flux at spectral locations in the tail or wings of the band increases with increasing optical depth and overcomes any decrease in spectral flux at the band head. The spectral location of the maximum spectral flux shifts farther and farther into the tail or wings of the band as the optical depth is increased. Only when the optical depth is enormous and adjacent bands overlap or the bands become so wide that $B(\nu,T_g)$ can no longer be approximated by $B(\nu_k,T_g)$ will a maximum in total flux be reached with increasing density. Thus a cold boundary layer does not completely shield the wall from hot molecular gas radiation.

7.F(f) Wall Layer Transmissivity. More insight into the effect of a cold boundary layer can be gained by defining a wall layer transmissivity. For a single spectral wavelength

$$\tau_{WL} = \frac{\varepsilon_{channel}[B_{g,center}-B_W]}{[1-2E_3(t_L)][B_{g,volume}-B_W]} \tag{7-76}$$

With the assumed trapezoidal profile in B, the wall layer transmissivity is

$$\tau_{WL} = \frac{\varepsilon_{channel}(t_L,t_{BL})}{[1-2E_3(t_L)][1-(t_{BL}/t_L)]} \tag{7-77}$$

In contrast, for an entire exponential-tailed band

$$\tau_{WL} = \frac{C^*(t_L, t_{BL})}{A_s^*(t_L)[1-(t_{BL}/t_L)]} \tag{7-78}$$

Table 7-2 shows the τ_{WL} behaviors of Eq. (7-77) and (7-78). The spectral value at the band head drops rapidly with increasing absorber density as t_{BL} grows past unity while the integrated band value falls more slowly as $1/\ell n(t_L)$ instead of $1/t_{BL}$.

One practical consequence of this behavior is that if one scales up the size of a combustion chamber or furnace under nonluminous combustion conditions, one expects the radiant heat flux at the wall to increase somewhat. Furthermore, regions of hot gas at distances far removed from the wall can be radiatively cooled by the cold wall because the optical depth in the band wings is not great, and the increase of $\tilde{\omega}_k$ in hot regions is marked. In contrast, in highly sooty combustion conditions the wall heat flux may decrease with scale up in size, and regions of flame remote from the wall cannot "see" the wall and thus receive no radiative cooling. This latter effect can lead to increased nitric oxide pollutant formation.

REFERENCES TO SECTION 7

I. Geometrical Considerations

A. The Slab

1. Kourganoff, V., Basic Methods in Transfer Problems, Oxford, London, 1952.

B. The Cylinder

2. Kestin, A. S., "Radiant Heat Flux Distribution in a Cylindrically Symmetric Nonisothermal Gas with Temperature Dependent Absorption Coefficients", J. Quantitative Spectroscopy and Radiative Transfer, Vol. 8, pp. 419-434, 1968.

3. Wassel, A. T., and D. K. Edwards, "Molecular Gas Band Radiation in Cylinders", J. Heat Transfer, Vol. 96, pp. 21-26, 1974.

II. Differential Formulations

A. Radiation Diffusion

4. Deissler, R. G., "Diffusion Approximation for Thermal Radiation in Gases with Jump Boundary Conditions", J. Heat Transfer, Vol. 86, pp. 240-246, 1964.

B. Multiflux Methods

5. Larkin, B. K., and S. W. Churchill, "Heat Transfer by Radiation Through Porous Insulations", A.I.Ch.E. Journal, Vol. 5, pp. 467-474, 1959.

6. Incropera, B. K., and W. G. Houf, "A Three-Flux Method for Predicting Radiative Transfer in Aqueous Suspensions", J. Heat Transfer, Vol. 101, pp. 497-501, 1979.

7. Modest, M. F., "A Simple Differential Approximation for Radiative Transfer in Non-Gray Gases", J. Heat Transfer, Vol. 101, pp. 735-736, 1979.

C. Embedding

8. Bellman, R. E., R. E. Kalaba, and M. C. Prestrud, *Invariant Imbedding and Radiative Transfer in Slabs of Finite Thickness*, American Elsevier, New York, 1963.

9. Rogers, J. E., and D. K. Edwards, "Bidirectional Reflectance and Transmittance of a Scattering-Absorbing Medium with Rough Surface", *Progress in Astronautics and Aeronautics*, Vol. 49, pp. 3-24, 1976.

D. Other Methods

10. Ozisik, M. N., *Radiative Transfer*, Wiley, New York, 1973, Chapters 9 and 10.

III. Spectral Considerations

A. Narrow Band Scaling

11. Goody, R. M., *Atmospheric Radiation*, Oxford, London, 1964, pp. 234-243.

12. Tiwari, S. N., and S. K. Gupta, "Accurate Spectral Modeling for Infrared Radiation", *J. Heat Transfer*, Vol. 100, pp. 240-246, 1978.

B. Wide Band Scaling

13. Edwards, D. K., and S. J. Morizumi, "Scaling of Vibration-Rotation Band Parameters for Nonhomogeneous Gas Radiation", *J. Quantitative Spectroscopy and Radiative Transfer*, Vol. 10, pp. 175-188, 1970.

14. Edwards, D. K., and A. Balakrishnan, "Slab Band Absorptance for Molecular Gas Radiation", *J. Quantitative Spectroscopy and Radiative Transfer*, Vol. 12, pp. 1379-1387, 1972. See also "Self-Absorption of Radiation in Turbulent Molecular Gases", *Combustion and Flame*, Vol. 20, pp. 401-417, 1973.

EXERCISES FOR SECTION 7

1. An absorbing-emitting gas has a temperature distribution such that its black body radiosity varies with distance from a wall as $B_o + (B_\infty - B_o)(1-e^{-y/\delta})$. (a) What is the intensity of the irradiation on the wall as a function of polar angle θ? (b) What is the incident flux? (c) Find a numerical value for q^- for the case of a gray gas with $k = 1\ cm^{-1}$, $\delta = 1$ cm, $T_o = 500$ K and $T_\infty = 2000$ K.

2. A gray absorbing-emitting gas with an absorption coefficient of $0.1\ cm^{-1}$ is contained between two infinite black parallel plates 7 cm apart, one on top at 2000 K and the other on the bottom at 1000 K. A 4 cm layer within the gas is stirred so it is isothermal at a temperature T_g corresponding to radiative equilibrium. The black body radiosity of the gas in the 1.0 cm thick layer above and in the 2 cm thick layer below the 4 cm mixed zone varies linearly from B_g to the wall values. Find the temperature T_g of the mixed-zone gas.

3. A furnace is modeled crudely as two infinite parallel black walls 0.9 m apart. The central 0.3 m is a flame at 2000 K with $k = 4\ m^{-1}$. The upper wall is at 1850 K, and the bottom wall at 1000 K. The top and bottom gas layers have $k = 0.04\ m^{-1}$ and are at 1400 K. (a) What

3. (continued)

is the heat flux into the bottom wall? (b) Numbering from the lower wall through the three gas zones to the upper wall from 1 to 5, find the values of the elements of the 5 x 5 matrix for $\mathcal{G}_{ij}$ per a unit area on one of the walls.

4. Consider an *optically thin* gas between two infinite parallel plates, one plate with temperature T_1 and emissivity ε_1 and the other with temperature T_2 and emissivity $\varepsilon_2 = 1$. The net volumetric heat source due to radiation as a function of location z, $0 < z < L$, is desired, assuming T(z) is known. (a) Write the equation for $\dot{Q}_V$. (b) Provide a numerical value for $\dot{Q}_V$ for L = 4 cm, $T_1 = 1000$ K, $T_2 = 500$ K, $\varepsilon_1 = 0.3$ and $k_a = 0.01\ \text{cm}^{-1}$ at a location where T is 800 K. Why is the value of z at which T(z) = 800 K immaterial?

5. Repeat Exercise 4 for two concentric spheres, $R_1 = 4$ cm and $R_2 = 8$ cm. Find $\dot{Q}_V$ at location r = 5 cm.

6. Repeat Exercise 4 for two concentric infinite cylinders $R_1 = 4$ cm and $R_2 = 8$ cm. Find $\dot{Q}_V$ at location r = 5 cm.

7. A slab of fused silica is 3 mm thick and contacts a black material (also with n = 1.5) at 1400°C on one side and another such material at 200°C on the other side. Sup-

7. (continued)

pose that the temperature variation through the slab is linear and that the optical properties of the silica can be approximated as follows: $n = 1.5$, $4\pi k/\lambda = k_a = C(\lambda/\lambda_o)^m$ where $C = 0.4\ mm^{-1}$, $\lambda_o = 1\ \mu m$, and $m = 1$.

(a) Find the net heat flux due to radiation at the 200°C surface.

(b) Find the heat flux, if, by adding pigment to the silica, C can be raised to $4\ mm^{-1}$ and m reduced to zero. Can you use the Rosseland diffusion approximation?

(Don't forget the n^2 multiplying the Planck formula; recall Exercise 8(b) of Section 2.)

8. Fibrous insulation consists of 75 µm diameter black opaque cylinders, all parallel to the plane $z = 0$ but otherwise randomly oriented. The material from which the fibers are made has a density of 2400 kg/m^3 and the fibrous insulation has a density of 40 kg/m^3. Estimate the radiation contribution to the thermal conductivity.

9. A beam of laser radiation of wavelength λ_o gives an irradiation of $I_o \Delta\Omega_o \cos\theta_o \Delta\lambda_o$. The irradiation is partly absorbed and partly transmitted by a plasma of thickness δ located next to a wall. The wall has a reflectance at λ_o of ρ_o. The plasma has absorption coefficient k_o at wavelength λ_o and initially $k_o\delta << 1$. The plasma is

9. (continued)

heated by the laser to a temperature so high that the radiation emitted by the plasma is at very short wavelengths. In the short wavelengths the plasma absorption coefficient is so large that the Rosseland diffusion approximation may be invoked, $q_R = -(16/3K_R)\sigma T^3(dT/dy)$. The heat absorbed from the laser flows according to the Rosseland diffusion approximation toward both boundaries, the wall and the edge of the plasma distance δ away from the wall. Both boundaries are so cold relative to the plasma that they may be assumed to be near zero in temperature. Answer the following questions:

(a) What is the fraction of laser radiation absorbed by the plasma for arbitrary values of $k_o\delta$?

(b) What volume heat source as a function of distance from the wall is caused by absorption of laser radiation for arbitrary values of $k_o\delta$?

(c) In the limit as $k_o\delta$ is small what is the volume heat source?

(d) Assuming negligible heat storage within the plasma, what fraction of the absorbed laser radiation is delivered to the wall when (i) $k_o\delta$ is small and (ii) when it isn't?

10. Find the heat transfer through an evacuated space 2.5 cm thick filled with a porous insulation with k_a = 160

10. (continued)

m^{-1}. The hot wall is at 15°C and the cold wall at 5°C. (a) Both walls are black. (b) Both walls have an emissivity of 0.03. (c) Compare your answer for part (b) with the flux between two walls with $\varepsilon = 0.03$ containing no insulation.

11. Formulate the bidirectional transmissivity of a finite slab using embedding. Demonstrate Helmholtz reciprocity by comparing results from embedding from below with embedding from above.

12. A column of nonisothermal gas has a temperature gradient such that the normalized black body intensity at wavenumbers near ν_o varies linearly from 0.1 at the cold end to 1.0 at the hot end. The gas possesses an exponential-tailed absorption band in the spectral vicinity of ν_o such that $k_a = k_o \exp[-(\nu_o - \nu)/\omega]$. The gas density and the column length are such that $k_o L = 100$. (a) Find the spectral absorptivity of the gas column as a function of $\nu^* = (\nu_o - \nu)/\omega$ for the interval $0 \leq \nu^* \leq 10$. (b) Find the intensity out of the hot end versus ν^* in the same interval. (c) Find the intensity out of the cold end vs. ν^*. (d) If the gas in question is CO_2 and $\nu_o = 2400\ cm^{-1}$, and the hot end temperature is 2000 K, find the cold end temperature and the total cold end intensity (W/m^2 sr) for the entire band.

13. Carbon monoxide gas at 10 atm pressure and 1500 K flows between two black parallel plates 10 cm apart. The walls are at 1000 K. Find the net heat flux at a wall when (a) the gas is completely isothermal (the boundary layer thickness is negligible); (b) a 1.0 cm thick boundary layer exists at each wall and the temperature distribution in the boundary layer is such that the Planck function at 2143 cm^{-1} rises linearly from the 500 K wall value to the 1500 K channel core value.

8 RADIATION ACTING WITH CONDUCTION OR CONVECTION

8.A Combined Phenomena. Radiation heat fluxes are linear in blackbody radiosity B on a spectral basis, and when spectral property variations are not severe this linearity carries over to the total radiosity $B_T = \sigma T^4$. Conduction heat fluxes are linear in heat flux potential $\Phi = \int K dT$ and are thus linear in temperature T when thermal conductivity K is constant. Convection is likewise linear in T when property variations are secondary. The result is that a linear problem in conduction or convection is generally made into a nonlinear problem when radiation is added. Closed-form analytical solutions are thus rare, and numerical calculations are commonly required. Even when a closed-form result is not derivable, putting the nonlinear problem into dimensionless form may disclose that the number of parameters is sufficiently small so that numerical calculation of general results can be made.

In special circumstances linearization is an appropriate procedure. With the radiation linearized in T (or more rarely the conduction or convection linearized in B) a closed-form result may be obtainable, and such a result even if not directly useful may convey insight to sharpen design intuition. Even when a problem is strongly nonlinear, quasilinearization will prove useful in constructing a solution procedure.

In what follows the simpler combined problems are considered, ones for which a closed form solution exists or ones which are parameterized in a small number of parameters. The linearization procedure and the useful concepts of a radiation heat transfer coefficient and equivalent environmental temperature are described.

8.B <u>Thermal Network Analysis</u>. Consider the thermal analysis of a simple system involving both radiation and convection. For example, consider a small room whose exterior wall of 12 m^2 is shaded single-glazed window and whose interior walls, ceiling, and floor of 60 m^2 are roughly adiabatic. The room contains heat sources of 1 kW, and the outside air is 30°C. The room air is cooled to 22°C by supply air at 12°C. It is agreed to calculate convective heat transfer coefficients on all surfaces with the approximate relation $h_c = 1.3\Delta T^{1/3}$ W/m^2 °C. Suppose the outside surface surrounds are infrared-black and are solar heated to 50°C, the shape factor from the outside of the window to the surrounds being 0.5 and the remainder to the sky. The outside air is at 60% relative humidity. The thermal designer wants to know how much cold air must be supplied to the room, and what energy savings there might be from installation of double-glazed window and/or insulated exterior wall.

The problem can be divided into three segments (i) radiation analysis of the exterior, (ii) radiation analysis of the interior, and (iii) thermal analysis of the system. These three segments illustrate a number of features of elementary thermal system analysis with combined phenomena: (1) definition of the mean radiant temperature, (2) the radiation heat transfer coefficient, (3) definition of the mean equivalent temperature, (4) the distinction between the radiation network and the thermal network, and (5) convergence problems.

8.B(a) Radiation Analysis of the Exterior. Under the assumption that the radiosity of the exterior surface has negligible influence upon the radiosity of the surrounds, the net radiant heat flux into the exterior surface 1 is

$$q_{r,o} = \alpha_{IR} q_{IR}^{-} + \alpha_S q_S^{-} - \varepsilon_1 \sigma T_1^4 \qquad (8\text{-}1)$$

Note that the glass is treated as an opaque surface, because its characteristic wavelength for cut-off of transmission is approximately $\lambda_{CO} = 2.7\ \mu m$, and $\lambda_{CO} T \doteq 810\ \mu m\ °K$ in Table 1-1 indicates a negligible internal fraction at wavelengths shorter than λ_{CO}. We begin by making the approximation $\alpha_{IR} \doteq \varepsilon_1 \doteq 0.82$ and calculating the infrared irradiation

$$q_{IR}^{-} = F_{1-s} q_s^{-} + F_{1-sky} q_{sky}^{+} \qquad (8\text{-}2)$$

According to the problem statement $q_s^{+} = \sigma T_s^4$ and F_{1-s} and F_{1-sky} are known from the geometry. The sky radiosity is

$$q_{sky}^{+} = \varepsilon_{sky} \sigma T_o^4 \qquad (8\text{-}3)$$

where ε_{sky} for the atmospheric CO_2, H_2O and dust is given approximately by the Brunt equation

$$\varepsilon_{sky} \doteq 0.55+1.8\sqrt{P_{H_2O}/P_0} \leq 1\ ,\quad P_0 = 1\text{ atm} \tag{8-4}$$

Since the partial pressure of H_2O is given by the product of the relative humidity and vapor pressure of water at the outside air temperature T_o, we have with the aid of a steam table $\varepsilon_{sky} = 0.836$ and $q^+_{sky} = 401\ W/m^2$. From Eq. (8-2) $q^-_{IR} = 509\ W/m^2$. According to the problem statement the solar irradiation is zero; thus all terms in Eq. (8-1) except the unknown T_1 are available.

The next step may be to calculate the mean radiant temperature of the surrounds, defined by

$$q_{r,o} = \varepsilon_1(\sigma T_r^{\ 4} - \sigma T_1^{\ 4}) \tag{8-5}$$

so by comparison with Eq. (8-1)

$$\sigma T_r^{\ 4} = \frac{\alpha_{IR}q^-_{IR} + \alpha_S q^-_S}{\varepsilon_1} \tag{8-6}$$

For the example at hand, $T_r = 307.9$ K (34.7°C). Here T_r is somewhat above T_o because of the shape factor to the hot T_s.

8.B(b) <u>The Radiation Heat Transfer Coefficient</u>. The next step may be to write Eq. (8-5) in the form

$$q_{r,o} = h_{r,o}(T_r - T_o) \tag{8-7}$$

where $h_{r,o}$ is the radiation heat transfer coefficient. By factoring Eq. (8-5)

$$h_{r,o} = \varepsilon_1\sigma(T_r^{\ 2}+T_1^{\ 2})(T_r+T_1) \tag{8-8}$$

Note that the value depends upon the unknown temperature T_1.

In the example here, common experience of touching a window under the conditions cited suggests $T_1 \doteq (T_i+T_o)/2 = 26°C$. Thus $h_{r,o} \doteq 5.20\ W/m^2\ °C$. Note that it is essential to use absolute temperatures (K) in Eq. (8-8), but once $h_{r,o}$ is known, relative temperatures can be used in Eq. (8-7).

In general, when we have

$$\dot{Q}_{r,i-j} = A_i \mathscr{F}_{i-j}(\sigma T_i^4 - \sigma T_j^4) \tag{8-9}$$

we can, if it suits our purposes, write

$$\dot{Q}_{r,i-j} = A_i h_{r,i-j}(T_i - T_j) \tag{8-10}$$

where

$$h_{r,i-j} = \mathscr{F}_{i-j}\sigma(T_i^2+T_j^2)(T_i+T_j) \tag{8-11}$$

8.B(c) The Mean Equivalent Temperature. It is convenient to consider the sum of the convective and radiative heat transfer into the exterior surface as coming from a single equivalent source. Accordingly one writes

$$q_{1,o} = h_{c,o}(T_o-T_1) + h_{r,o}(T_r-T_1) = h_o(T_e-T_1) \tag{8-12}$$

Again with $h_{c,o} \doteq 1.3(T_o-T_1)^{1/3}$ (say) the value depends upon T_1. We use our estimate of $T_1 \doteq 26°C$ to find $h_{c,o} \doteq 2.06\ W/m^2\ °C$. From inspection of Eq. (8-12)

$$h_o = h_{c,o} + h_{r,o} \tag{8-13}$$

$$T_e = \frac{h_{c,o}T_o + h_{r,o}T_r}{h_{c,o}+h_{r,o}} \tag{8-14}$$

With the estimates we find $h_o = 7.26\ W/m^2\ °C$ and $T_e = 33.4°C$, the latter somewhat above T_o because of the hot value of T_s.

8.B(d) Radiation Analysis of the Interior. Let node 2 be the inside of the exterior wall, node 3 the remaining wall surface and node 4 the interior gas. The first step is to make a radiation analysis to find the transfer factors. Neglecting the gas emissivity (the reader should redo the more intricate problem when the gas is absorbing-emitting), we have from a radiation network

$$A_2\mathscr{F}_{2-3} = \frac{1}{\dfrac{1-\varepsilon_2}{\varepsilon_2 A_2} + \dfrac{1}{A_2 F_{2-3}} + \dfrac{1-\varepsilon_3}{\varepsilon_3 A_3}} \tag{8-15}$$

For the values given, and assuming $\varepsilon_2 = 0.82$, $\varepsilon_3 = 0.87$ we obtain $\mathscr{F}_{2-3} = 0.800$. The radiation heat transfer coefficient $h_{r,2-3}$ is then found from Eq. (8-11) using estimated values of $T_2 = 26°C$ (299.15 K) and $T_3 = 24°C$ (297.15 K), $h_{r,2-3} \doteq 4.81$ W/m^2 °C.

8.B(e) Construction and Solution of the Thermal Network. The thermal network is now readily constructed as shown in Fig. 8-1. The exterior environment at equivalent temperature T_e drives heat through outside thermal resistance $1/h_oA_1$ by convection and radiation, through the exterior wall by conduction, from the inside of the exterior wall to the air directly by convection and indirectly by radiation to the interior walls and then convection to the room air. Note that the radiation

FIGURE 8-1

Thermal Network for a Room

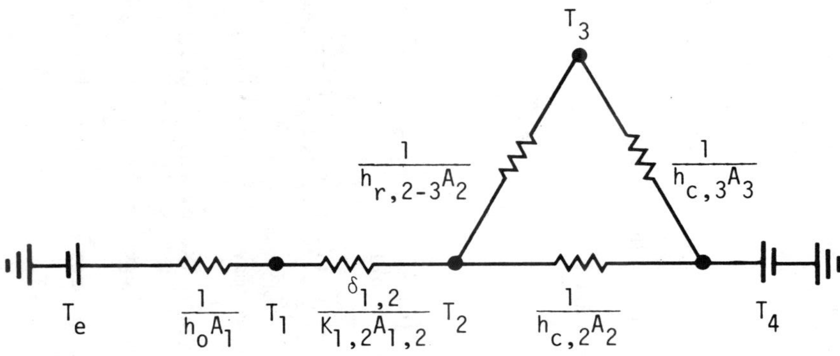

network is a prelude to the thermal network. The thermal network includes both radiation and convection resistances, both nonlinear.

The heat transfer is given by

$$\dot{Q} = \frac{T_e - T_4}{\dfrac{1}{h_o A_1} + \dfrac{\delta_{1,2}}{K_{1,2}A_{1,2}} + \dfrac{1}{h_{c,2}A_2 + \dfrac{1}{\dfrac{1}{h_{r,2-3}A_2} + \dfrac{1}{h_{c,3}A_3}}}} \tag{8-16}$$

Note that the order of the subscripts in the $A_i h_{r,i-j}$ terms must correspond to the order in the $A_i \mathscr{F}_{i-j}$ term upon which the former is based.

The unknown temperatures T_1, T_2, and T_3 follow from the network

$$T_1 = T_e - \dot{Q}\left[\frac{1}{h_o A_1}\right], \quad T_2 = T_e - \dot{Q}\left[\frac{1}{h_o A_1} + \frac{\delta_{1,2}}{K_{1,2}A_{1,2}}\right] \tag{8-17a,b}$$

$$T_3 = T_4 + [\dot{Q} - (T_2 - T_4)h_{c,2}A_2]\left[\frac{1}{h_{c,3}A_3}\right] \tag{8-17c}$$

To solve we may execute a simple relaxation scheme; however, convergence is not guaranteed, but in this example there is no problem. We estimate $h_{c,2}$ as 2.06 W/m^2 °C and $h_{c,3}$ as 1.64 W/m^2 °C and for soda-lime glass 3 mm thick $\delta_{1,2}/K_{1,2}$ = 0.0034 (W/m^2 °C)$^{-1}$, a negligible value. Using these values we obtain $\dot{Q}$ = 405 W from Eq. (8-16), and find T_1, T_2, and T_3. Then we return to the defining equations and refine T_e, h_o, $h_{r,2-3}$, $h_{c,2}$ and $h_{c,3}$ and repeat the process until convergence is attained. Note that our assumed problem had no outside forced convection and no leakage of outside air into the room, hence the low value of $\dot{Q}$.

The rate of air circulation required is given by the First Law of Thermodynamics written for the open flow room air system

$$\dot{m}c_p(T_4 - T_{4,i}) = \dot{Q} \tag{8-18}$$

Taking c_p = 1006 J/kg °C and $T_{4,i}$ = 12°C, we find $\dot{m}$ = 0.040 kg/s for the 405 W, and $\dot{m}$ = 0.140 kg/s for 405 W plus 1 kW.

Convergence in the above procedure may not be achieved in a simple relaxation scheme, because an underprediction of an h_r may lead to an overprediction of a T, which may then lead to an overprediction of the h_r followed by an underprediction of the T with no closure. Convergence may be achieved by using a damping factor d, i.e., the new estimate of T_1 is taken to be the last estimate times (1-d) plus d times the newly calculated value. Alternatively one may try constructing a thermal network (still distinct from the radiation network used to get the transfer factors) in which the driving potentials are $B = \sigma T^4$ instead of T, and instead of using h_r one uses the $A_i \mathscr{F}_{i-j}$ factors directly and replaces the $1/h_{c,i}A_i$ resistances with $1/A_i \mathscr{F}_{c,i-j}$ where

$$A_i \mathscr{F}_{c,i-j}(\sigma T_i^4 - \sigma T_j^4) \equiv A_i h_{c,i}(T_i - T_j)$$

$$\mathscr{F}_{c,i-j} = \frac{h_{c,i}}{\sigma(T_i^2 + T_j^2)(T_i + T_j)} \tag{8-19}$$

8.C Radiation-Coupled Transient Heating or Cooling. Consider a mass m with specific heat c initially at temperature T_o suddenly exposed to a radiant environment at temperature T_r via an area A and transfer factor $\mathcal{F}$ and simultaneously heated (+) or cooled (-) internally at the rate $\dot{Q}$. If the internal resistance due to conduction (or convection if the mass is liquid) is small compared to the surface radiation resistance, i.e., if $\delta/KA << 1/A\mathcal{F}4\sigma T^4$, where δ is an appropriate length for internal conduction and K is the internal thermal conductivity, then the mass remains sufficiently isothermal to write

$$mc\,\frac{dT}{dt} = \dot{Q} - A\mathcal{F}(\sigma T^4 - \sigma T_r^{\,4}) \qquad (8\text{-}20)$$

The equation is simplified into dimensionless form by writing

$$T_e = [T_r + \dot{Q}/A\mathcal{F}\sigma]^{1/4}\ , \quad \beta = T_e/T_o \qquad (8\text{-}21a,b)$$

$$T^* = T/T_o\ , \quad t^* = t/t_c \qquad (8\text{-}21c,d)$$

$$t_c = mc/(A\mathcal{F}\sigma T_o^{\,3}) \qquad (8\text{-}21e)$$

The dimensionless form of Eq. (8-20)

$$\frac{dT^*}{dt^*} = -(T^{*4} - \beta^4) \qquad (8\text{-}22)$$

is separable and integrable into closed form for t* vs T* [2]

$$t^* = \frac{1}{2\beta^3}\left\{\frac{1}{2}\,\ell n\,\frac{(\beta+T^*)(\beta-1)}{(\beta-T^*)(\beta+1)} + \tan^{-1}\frac{T^*-1}{\beta+(T^*/\beta)}\right\} \qquad (8\text{-}23)$$

The solution obviously applies to a spacecraft radiator whose power is suddenly switched on or off, to a hot casting which is suddenly quenched by radiation, or to a cold object which is suddenly placed into a hot radiant furnace. One calculates T_e, β, and T_c via Eq. (8-21) and knowing the desired T or guessing the value, finds T* from Eq. (8-21c) and then t* from Eq. (8-23) and finally t from Eq. (8-21d).

If convection acts also, a linear term in T appears on the right side of Eq. (8-20), and the equation still separates. The required integral appears subject to closed-form integration but is tedious. If the convection is slight, one can write it in B according to Eq. (8-19) and modify $A\mathscr{F}$ and T_e accordingly. If the convection is dominant, one could introduce h_r. For a practical calculation, numerical integration can be readily carried out to obtain a table of t* vs T* with parameters β and $h^* = h_c/\mathscr{F}\sigma T_o^3$.

8.D Radiant Heat Exchanger. Suppose a fluid at temperature T_o with specific heat c_p flows at a mass flow rate of $\dot{m}$ through a tube or passage of total length L exposed on perimeter P with transfer factor $\mathscr{F}$ to a radiator at temperature T_r. If the overall heat transfer is radiation-controlled, i.e., $1/\mathscr{F}4\sigma T^3 \gg \delta_w/K_w$ or $1/h_i$ where δ_w is the tube wall thickness and K_w is its conductivity and h_i is the inside heat transfer coefficient, then the equation giving the temperature change with direction of flow z is

$$\dot{m}c_p \frac{dT}{dz} = -P\mathscr{F}(\sigma T^4 - \sigma T_r^4) \tag{8-24}$$

Using Eq. (8-19) one can make allowance for h_i and δ_w/K_w by finding equivalent transfer factors and summing the radiation resistances. Equation (8-24) is seen to be in the form of Eq. (8-20). Hence the solution obtained can be adopted. Let

$$\beta = T_r/T_o \ , \quad T^* = T/T_o \tag{8-25a,b}$$

$$z^* = PL\mathscr{F}\sigma T_o^3/\dot{m}c_p \tag{8-25c}$$

Then $z^* = t^*$ is given by Eq. (8-23).

The form of Eq. (8-23) giving z^* versus T^* is that desired for design. It tells the designer the length of exchanger needed to achieve a desired outlet temperature. In conventional heat exchanger design terminology, z^* is the number of transfer units, and T^* is related to the effectiveness ε_{eff} by

$$T^* = 1+(\beta^*-1)\varepsilon_{eff} \tag{8-26}$$

where ε_{eff} has its usual meaning

$$\varepsilon_{eff} = \frac{T_{out}-T_o}{T_r-T_o} \tag{8-27}$$

For example, how much area is needed for a spacecraft waste heat exchanger to cool a stream having $\dot{m}c_p$ = 100 W/K from 450 K to 350 K? The exchanger is shaded from the sun, is remote from planetary bodies, but receives 100 W/m^2 irradiation from a sun shield. The transfer factor to the surrounds is 0.90. One begins by finding $T_r = [100/5.67\text{x}10^{-8}]^{1/4}$ = 205 K; hence β = 205/450 = 0.455. The desired value of T^* is 350/450 = 0.778. From Eq. (8-23) $t^* = z^*$ = 0.407, hence from Eq. (8-25c) $PL = [(100)(0.407)]/[0.90(5.67\text{x}10^{-8})(450^3)] = 8.76\ m^2$.

8.E The Radiating Fin. In the example just described it was tacitly assumed that every element of perimeter was fully effective. In actuality one might use the skin of the vehicle for the radiator surface with tubes carrying the fluid stream spaced distance 2ℓ appart as sketched in Figure 8-2. Such an arrangement provides the fluid conduits some micrometeorite protection. While it would be reasonable to assume that the segment of perimeter ℓ_o is isothermal, the ℓ segments of perimeter on either side of the conduits are clearly noniso- thermal when b is small.

Let x be the distance from the midpoint between conduits in the ℓ direction. Let the radiating area of a dx element be Pdx (in Fig. 8-2 P is perpendicular to the cross-section shown and could be called Δz). Let the conduction cross-section be A_c (= $b\Delta z$ in Fig. 8-2). The transfer factor to the surrounds is called ε, and the equivalent radiant temperature, given by Eq. (8-6), is T_r. At $x = \ell$ the base temperature is T_B, and at $x = 0$, by symmetry, $dT/dx = 0$. For constant conductivity K of the material in area A_c the governing equation is

$$KA_c \frac{d^2T}{dx^2} = \varepsilon P\sigma(T^4 - T_r^{\ 4}) \qquad (8\text{-}28)$$

Here the fin approximation, that the surface temperature and

FIGURE 8-2

A Radiating Fin Configuration

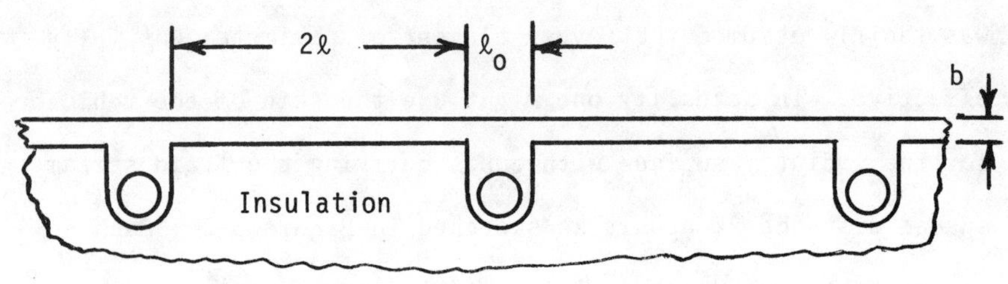

FIGURE 8-3

Fin Effectiveness for the Simple Radiating Fin [5]

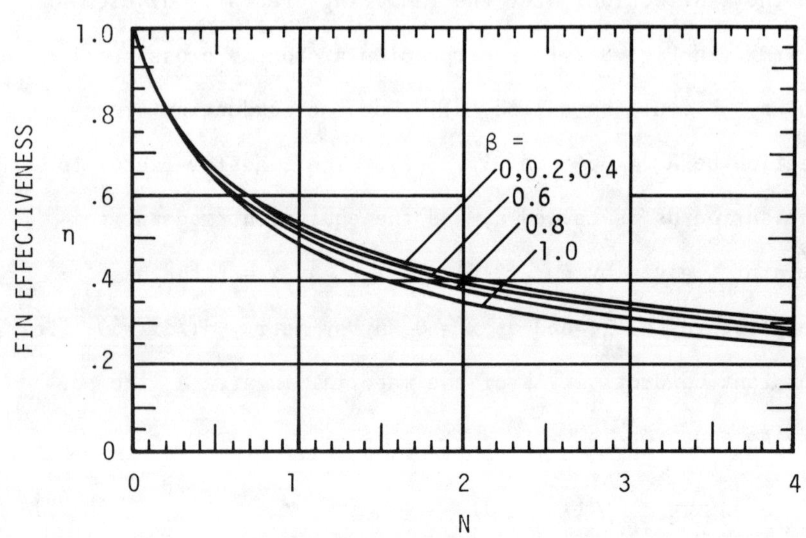

area A_c average temperature are indistinguishable, has been made. In dimensionless form, one has

$$\frac{d^2T^*}{dx^{*2}} = N(T^{*4}-\beta^4) \tag{8-29}$$

subject to

$$x^* = 0 \ , \quad dT^*/dx^* = 0 \ ; \quad x^* = 1 \ , \quad T^* = 1$$

where

$$T^* = T/T_B \ , \quad \beta = T_r/T_B \ , \quad x^* = x/\ell \tag{8-30a,b,c}$$

$$N = \varepsilon P\sigma T_B{}^3\ell^2/KA_c = \varepsilon\sigma T_B{}^3\ell^2/Kb \tag{8-30d}$$

The parameter N is like a radiation Biot number $h_r b/K$ times the square of a slenderness ratio ℓ/b.

Fin effectiveness η is defined as

$$\eta \equiv \frac{\int_0^\ell \varepsilon P\sigma(T^4-T_r{}^4)\,dx}{\varepsilon P\sigma(T_B{}^4-T_r{}^4)\ell} \tag{8-31a}$$

$$\eta = \int_0^1 [(T^{*4}-\beta^4)/(1-\beta^4)]\,dx^* \tag{8-31b}$$

From a heat balance η is also related to dT^*/dx^* at $x^* = 1$.

It is clear that η is a function of N and β. Numerical calculations can be made readily by rescaling with respect to the temperature T_o at $x = 0$, i.e., replacing T^* with $\theta = T/T_o$ and β with $\theta_r = T_r/T_o$ in Eq. (8-29) and introducing $\Phi = d\theta/dx^*$. Then moving θ_r $(0 \leq \theta_r < 1)$ as an arbitrary parameter gives one an initial value problem, thus avoiding the split boundary condition, as follows:

$$x^* = 0 \ , \quad \theta = 1 \ (\theta_r \text{ assumed}) \ , \quad \Phi = 0$$

$$\frac{d\Phi}{dx^*} = N(\theta^4 - \theta_r^{\ 4}) \ , \quad \frac{d\theta}{dx^*} = \Phi$$

The two equations are solved simultaneously by a Runge-Kutta integration, keeping the running sum $S^* = \int \theta^4 dx^*$, to find $\theta = \theta_B$ at $x^* = 1$. With θ_B known, $\beta = \theta_r/\theta_B$ and

$$\eta = [(S^*(1)/\theta_B^{\ 4}) - \beta^4]/[1-\beta^4]$$

Thus η can be plotted versus β for desired values of N. Figure 8-3 shows a cross-plot of η versus N with parameter β from [5].

If, instead of the uniform thickness b shown in Fig. 8-2, the thickness b is tapered, a lighter fin can be achieved. With variable A_c, Eq. (8-28) becomes

$$K \frac{d}{dx} (A_c \frac{dT}{dx}) = \varepsilon P \sigma (T^4 - T_r^{\ 4}) \qquad (8\text{-}32)$$

With dT/dx taken constant at $(T_B - T_o)/\ell$, T is known as a function of x, and the designer can integrate to find how A_c should vary with x. Let

$$A^* = A/A_B = b/b_B, \quad T^* = T/T_B, \quad x^* = x/L, \quad \beta = T_r/T_B \qquad (8\text{-}33a\text{-}d)$$

$$N_B = \varepsilon P \sigma T_B^{\ 3} \ell^2/KA_B = \varepsilon \sigma T_B^{\ 3} \ell^2/Kb_B \qquad (8\text{-}33e)$$

Here it is convenient to regard T_o^* ($\beta < T_o^* < 1$) as a parameter. For a value of T_o^*, N_B is given by

$$N_B = \frac{5(1-T_o^*)^2}{(1-T_o^{*5})-5\beta^4(1-T_o^*)} \tag{8-34}$$

and

$$A^* = \frac{(T^{*5}-T_o^{*5})-5(T^*-T_o^*)}{(1-T_o^{*5})-5(1-T_o^*)} \tag{8-35}$$

where

$$T^* = T_o^* + (1-T_o^*)x^* \tag{8-36}$$

Fin effectiveness η is given by

$$\eta = \frac{(1-T_o^{*5})-5\beta^4(1-T_o^*)}{5(1-\beta^4)(1-T_o^*)} \tag{8-37}$$

For example, what thickness b_B is needed for a 90 per cent effective fin with K = 200 W/m K, ℓ = 5 cm, T_B = 450 K, and T_r = 225 K, and ε = 0.903, and how should b/b_B vary with x/ℓ? One begins by finding β = 0.50 and using Eq. (8-37) to find T_o^* = 0.95076. From Eq. (8-34) N_B = 0.05836. From Eq. (8-33e) b_B = 0.996 mm. The fin profile is given by Eq. (8-35) as follows:

x^*	b/b_B
1.00	1.00
0.80	0.958
0.60	0.835
0.40	0.633
0.20	0.354
0.0	0.0

Needless to say there are many permutations of the radiating fin problem. Sparrow and Cess [5] give a review of many of them. The next section considers (in the linearized limit) a fin that views itself so that the irradiation upon it depends upon the temperature distribution along the fin.

8.F Linearized Radiation and Conduction Around the Hollow Cylinder. Hollow cylinders are often used as booms or struts for space vehicles. When one is exposed to solar heating from one side it may bend into a bow shape from the greater thermal expansion of the hot side. The problem is taken up here to give an example of an exact solution of the diffuse-walled enclosure problem and to show how linearizing a combined radiation and conduction problem facilitates analysis. Sparrow [5] pointed out that the hollow circular cylinder was one of the few enclosures for which an exact solution of the radiosity problem can be readily found. Hrycak and Helgans [6] considered the hollow conducting cylinder with a black interior. Edwards [7] treated the hollow cylinder with gray diffuse interior and an anisotropically conducting wall.

Consider a hollow cylinder of wall thickness b and radius R having temperature $T(\theta)$ close to T_o and with b sufficiently thin so that $b/R \ll 1$ and $4\sigma T_o^3 b/K \ll 1$. The angle θ is measured from the solar stagnation line; $\theta = \pm\pi/2$ occurs on the shadow lines, and $\theta = \pi$ is on the dark side as pictured in Fig. 8-4. The fin equation is obtained from a heat balance on an element between θ and θ+dθ

$$\frac{Kb}{R^2}\frac{d^2T}{d\theta^2} = \varepsilon_o \sigma T^4 + q_i - \alpha_S G_S \cos\beta \cos\theta\, U \qquad (8\text{-}38)$$

where ε_o is the total hemispherical emissivity of the outside surface, q_i is the net internal heat flux by radiation, $\alpha_S(\theta)$ is the solar absorptivity, G_S is the solar irradiation at normal incidence $I_S \Delta\Omega_S$, and β is the angle of the solar ray from the base plane normal to the cylindrical axis (recall Fig. 1-3; the half angle subtended by the sun is assumed to be small), and U is the unit step function, unity in the sun ($0 \leq \theta < \pi/2$) and zero in the shade ($\pi/2 < \theta \leq \pi$).

The first step in simplifying the problem for solution is linearization. One writes

$$\sigma T^4 \doteq \sigma T_o{}^4 + 4\sigma T_o{}^3 (T - T_o) \qquad (8\text{-}39)$$

and introduces

$$T^* = (T - T_o)/T_o \qquad (8\text{-}40)$$

Eq. (8-38) becomes

$$K^* \frac{d^2T^*}{d\theta^2} = 4T^* + q_i^*(\theta) - G^*(\theta) \qquad (8\text{-}41)$$

where

$$K^* = Kb/R^2 \varepsilon_o \sigma T_o{}^3 \qquad (8\text{-}42a)$$

FIGURE 8-4

Hollow Conducting Cylinder with Gray Diffuse Interior
Exposed to External Irradiation
from [7]

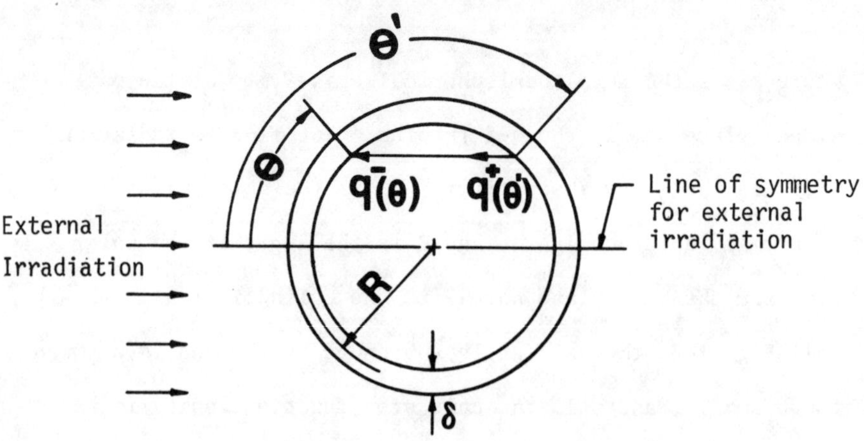

$$q_i^* = q_i/\varepsilon_o \sigma T_o^4 \tag{8-42b}$$

$$G^* = \frac{1}{\varepsilon_o \sigma T_o^4} [\alpha_S G_S \cos\beta \cos\theta \, U - \varepsilon_o \sigma T_o^4] \tag{8-42c}$$

To proceed, the internal enclosure problem is analyzed, subject to the assumption that the interior is diffuse with uniform internal total hemispherical emissivity ε_i. The irradiation is found via Eq. (3-8) in integral form

$$q^-(\theta) = \int_{-\pi}^{+\pi} q^+(\theta') dF(\theta,\theta')$$

where, from Eq. (1-15),

$$dF(\theta,\theta') = \frac{1}{2} \cos\gamma \, d\gamma$$

and γ is given by

$$\gamma = \frac{\pi}{2} - \frac{1}{2} |\theta'-\theta|$$

Thus the irradiation-radiosity equations are

$$q^-(\theta) = \int_{-\pi}^{\theta} q^+(\theta') \frac{1}{4} \sin(\frac{\theta-\theta'}{2}) d\theta' + \int_{\theta}^{\pi} q^+(\theta') \frac{1}{4} \sin(\frac{\theta'-\theta}{2}) d\theta' \tag{8-44a}$$

$$q^+(\theta) = \varepsilon_i \sigma T^4 + (1-\varepsilon_i) q^-(\theta)$$

$$q^+(\theta) = \varepsilon_i \sigma T_o^4 + 4\varepsilon_i \sigma T_o^4 T^* + (1-\varepsilon_i) q^-(\theta) \tag{8-44b}$$

The net transfer is

$$q_i = q^+ - q^- = \varepsilon_i \sigma T_o^4 [4T^* - q^*]$$

$$q_i^* = [\varepsilon_i/\varepsilon_o][4T^* - q^*] \tag{8-45}$$

where

$$q^* = (q^- - \sigma T_o^4)/\sigma T_o^4 \tag{8-46}$$

Substitution of Eq. (8-44b) into Eq. (8-44a) and introduction of Eq. (8-46) gives a linear integral equation in q* coupled through T* to the linear differential equation, Eq. (8-41).

$$q^*(\theta) = \int_{-\pi}^{\pi} [4\varepsilon_i T^*(\theta') + (1-\varepsilon_i) q^*(\theta')] \frac{1}{4} \sin\left(\frac{|\theta-\theta'|}{2}\right) d\theta' \tag{8-47}$$

A series solution is obtained by expanding G* in a Fourier cosine series

$$G^* = \sum_{n=1}^{\infty} a_n \cos n\theta \tag{8-48}$$

where T_o is chosen to make a_o be zero.

$$\sigma T_o^4 = \frac{1}{\pi\varepsilon_o} \int_0^{\pi/2} \alpha_S G_S \cos\beta \cos\theta \, d\theta \tag{8-49}$$

The coefficients a_n are found by multiplying both sides of Eq. (8-48), written with m replacing n, by cos nθ and integrating from θ = 0 to π.

$$a_n = \frac{2}{\pi} \int_0^{\pi/2} \frac{\alpha_S G_S \cos\beta}{\varepsilon_o \sigma T_o^4} \cos\theta \cos n\theta \, d\theta - \frac{2}{\pi} \int_0^{\pi} \cos n\theta \, d\theta \tag{8-50}$$

The last term is clearly zero. For constant α_S (a diffuse exterior) Eqs. (8-49) and (8-50) give

$$\sigma T_o^4 = \frac{\alpha_S G_S \cos\beta}{\varepsilon_o \pi} \tag{8-49a}$$

$$a_1 = \pi/2 \;, \quad a_{2m} = \frac{2(-1)^{m+1}}{(2m)^2-1} \tag{8-50a}$$

$$a_3 = a_5 = a_7 \ldots = 0$$

Then it is clear that Eqs. (8-41) and (8-47) are satisfied by

$$T^* = \sum_{n=1}^{\infty} b_n \cos n\theta \tag{8-51}$$

$$q^* = \sum_{n=1}^{\infty} c_n \cos n\theta \tag{8-52}$$

when the coefficients b_n and c_n are made compatible with a_n. For convenience the integral S_n is defined

$$S_n = \frac{1}{\cos n\theta} \int_{-\pi}^{\pi} \frac{1}{4} \sin\left(\frac{|\theta-\theta'|}{2}\right) \cos n\theta' d\theta' \tag{8-53}$$

$$S_n = -1/(4n^2-1) \tag{8-54}$$

Equation (8-47) then gives

$$c_n = 4\varepsilon_i S_n b_n + (1-\varepsilon_i) S_n c_n$$

$$c_n = \frac{4\varepsilon_i S_n}{1-(1-\varepsilon_i)S_n} b_n \tag{8-55}$$

and Eqs. (8-41,45, and 47) give

$$-K^* n^2 b_n = 4b_n + (\varepsilon_i/\varepsilon_o)[4b_n - c_n] - a_n$$

Substituting Eq. (8-55) and solving for b_n gives

$$b_n = \frac{1}{4\left[1 + \frac{\varepsilon_i}{\varepsilon_o}\left(\frac{1-S_n}{1-(1-\varepsilon_i)S_n}\right)\right] + n^2 K^*} a_n \tag{8-56}$$

The solution to the problem is now complete. In a practical engineering calculation two terms suffice. Hence the temperature is given by

$$T \doteq T_o[1+b_1\cos\theta + b_2\cos 2\theta] \qquad (8\text{-}51a)$$

where T_o is given by Eq. (8-49a) and

$$b_1 = \frac{\pi/2}{4[1 + \frac{4\varepsilon_i}{\varepsilon_o(4-\varepsilon_i)}] + \frac{Kb}{R^2\varepsilon_o\sigma T_o^3}} \qquad (8\text{-}56a)$$

$$b_2 = \frac{2/3}{4[1 + \frac{16\varepsilon_i}{\varepsilon_o(16-\varepsilon_i)}] + \frac{4Kb}{R^2\varepsilon_o\sigma T_o^3}} \qquad (8\text{-}56b)$$

The design implications of the above relations are clear. The maximum temperature at $\theta = 0$ minus the minimum temperature at $\theta = \pi$ is simply $2b_1T_o$. To minimize it one wants a small b_1. Hence a large ε_i and small ε_o are desirable, as is a large value of Kb/R^2 consistent with design weight and strength requirements. To use a small ε_o while maintaining a design goal for T_o requires adjusting α_S according to Eq. (8-49). As explained in Section 2.I, stripes of white paint or metallized tape may be used to trim the α_S/ε_o ratio.

For example, suppose $\varepsilon_i = 0.9$, $\varepsilon_o = 0.2$, $K = 50$ W/m K (graphite-reinforced phenolic), $b = 0.5$ mm, $R = 4$ cm, and $T_o = 300$ K. Equations (8-56a,b) give $b_1 = 0.0201$ and $b_2 = 0.00293$. Then Eq. (8-51a) gives $T(\theta=0) = 306.9$ K and $T(\theta=\pi) = 294.9$ K for the maximum and minimum temperatures respectively.

8.G Convection of a Radiating Molecular Gas. A final example is taken to illustrate how a combined-convection-radiation problem is analyzed. A molecular gas at inlet temperature T_o enters a section of plane parallel black duct with wall temperature T_w as shown in Fig. 8-5. The flow is assumed to be fully developed hydrodynamically; the radiation is linearized; the gas has a single exponential-tailed band (or a number of identical such bands) with overlapped lines; the duct walls are black. Balakrishnan and Edwards [19] analyzed this problem, and the nonlinear and multiband gas cases have also been examined. The general subject of coupled convection and radiation fills a fairly large literature, e.g. [10-19].

8.G(a) Formulation. The governing energy equation for low-speed steady, hydrodynamically-established, (negligible viscous dissipation and flow work) forced convection when axial conduction and radiation are neglected is

$$\rho c_p u \frac{\partial T}{\partial x} = \frac{\partial}{\partial y} [\rho c_p (\alpha_m + \varepsilon_H) \frac{\partial T}{\partial y}] + \frac{\partial}{\partial y} (-q_R) \quad (8\text{-}57)$$

where ρ is density, c_p specific heat at constant pressure, and α_m is molecular thermal diffusivity $K_m/\rho c_p$ (all at an appropriate reference temperature); u is velocity in the x-direction of flow; y is distance from one wall; ε_H is eddy diffusivity for heat; T(x,y) is gas temperature; and q_R is radiative heat flux in the y-direction. The velocity dis-

FIGURE 8-5

Plane-Parallel Black-Walled Duct
with Step Change in Wall Temperature

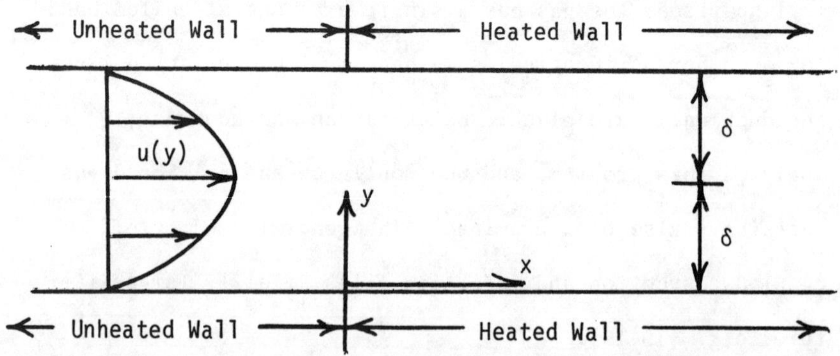

tribution u(y) under hydrodynamically-established conditions ($\partial u/\partial x \doteq 0$; hence $v \doteq 0$ from continuity) is

$$0 = -\frac{dP}{dx} + \frac{\partial}{\partial y}\left[\rho(\nu_m+\varepsilon_M)\frac{\partial u}{\partial y}\right] \qquad (8\text{-}58)$$

where ν_m is molecular kinematic viscosity and ε_M is eddy diffusivity for momentum. The pressure gradient is related to the skin friction coefficient c_f (or the Blasius friction factor $f = 4c_f$) by

$$-\frac{dP}{dx} = c_f \frac{1}{2} \rho U^2/\delta \qquad (8\text{-}59)$$

where U is the average velocity and δ the channel half width.

The eddy diffusivity for turbulent flow is taken to be given by

$$\nu_m+\varepsilon_M = \nu_m\varepsilon^+ \qquad (8\text{-}60)$$

where

$$\varepsilon^+ = 1 + \frac{1}{1+by^*}\left\{\frac{1}{2}\left(1+4K^2y^{+2}[1-e^{-y^+/A^+}]^2\right)^{1/2} - \frac{1}{2}\right\} \qquad (8\text{-}61)$$

and dimensionless distance is scaled with channel half width or turbulent mixing length

$$y^* = y/\delta\ , \quad y^+ = y^*R_t \qquad (8\text{-}62a,b)$$

Turbulent Reynolds number is

$$R_t = \delta^+ = (c_f/2)^{1/2}U\delta/\nu_m = (f/128)^{1/2}Re \qquad (8\text{-}63)$$

and is related as shown to Reynolds number based upon hydraulic diameter

$$Re = U(4\delta)/\nu_m \tag{8-64}$$

The Mei and Squire, von Karman, and Van Driest constants are

$$b = 3.4 \ , \quad K = 0.4 \ , \quad A^+ = 26 \tag{8-65a-c}$$

The eddy diffusivity for heat is taken to be given by

$$\alpha_m + \varepsilon_H = (\nu_m/Pr_m)(1+\varepsilon_M Pr_m/Pr_t) \tag{8-66}$$

where Pr_m is molecular Prandtl number ν_m/α_m and Pr_t is turbulent Prandtl number $\varepsilon_M/\varepsilon_H$. Radiation affects Pr_t somewhat [17], but the usually accepted value is approximately 0.9.

The radiant heat flux is written in terms of the slab band absorption by means of Eq. (7-70)

$$q_R = -\sum_{k=1}^{n} \omega_k B_k' \int_0^{\delta} K^*(y,y') \frac{\partial T}{\partial y'} dy' \tag{8-67}$$

where ω_k is band width parameter and B is the Planck function; $B_k' = \partial B(\nu_k,T)/\partial T$ at the reference temperature. Linearization of the radiation is thus assumed in factoring B_k' from under the integral. The kernel K* is given by Eq. (7-70) as

$$K^*(y,y') = A_s^*(t_H(2-y^*-y^{*\prime})) - A_s^*(t_H|y^*-y^{*\prime}|) \tag{8-68}$$

where A_s^* is given by Eq. (7-66) and t_H is the optical depth at the band head for half width δ. In terms of integrated band intensity α_k, the band head half depth is

$$t_H = \frac{\alpha_k \rho_a \delta}{\omega_k} \tag{8-69}$$

To proceed we make the governing equations dimensionless. Let

$$u^+ = u\delta/\nu_m R_t \ , \quad T^* = (T-T_w)/(T_o-T_w) \tag{8-70a,b}$$

$$R_{dm} = \frac{\delta}{K_m} \sum_{k=1}^{n} \omega_k B_k' A_s^* \tag{8-70c}$$

$$q_R^* = -q_R / \sum_{k=1}^{n} \omega_k B_k' A_s^* [T_o - T_w] \tag{8-70d}$$

where to shorten the notation

$$A_s^* = A_s^*(2t_H) \tag{8-70e}$$

and

$$x^* = x/(\delta Pr_m R_t) \tag{8-70f}$$

Note that the $A_s^*(2t_H)$ parameter is introduced as a convenience for later extracting a wall layer transmissivity τ_{WL} from the results. The dimensionless equations become

$$0 = R_t + \frac{d}{dy^*} (\varepsilon^+ \frac{du^+}{dy^*}) \tag{8-71}$$

$$u^+ \frac{\partial T^*}{\partial x^*} = \frac{\partial}{\partial y^*} (\varepsilon^+ \frac{\partial T^*}{\partial y^*}) + R_{dm} \frac{\partial}{\partial y^*} q_R^* \tag{8-72}$$

where

$$q_R^* = \sum_{k=1}^{n} \int_0^1 K^*(y^*,y^{*\prime}) \frac{\partial T^*}{\partial y^{*\prime}} dy^{*\prime}/A_k^* \tag{8-73}$$

Boundary conditions are $u^+ = 0$ and $T^* = 0$ at $y^* = 0$, du^+/dy^* and $dT^*/dy^* = 0$ at $y^* = 1$, and $T^* = 1$ at $x^* = 0$.

8.G(b) Numerical Solution. The momentum equation, Eq. (8-71) is solved by numerical integration

$$u^+ = R_t \int_0^{y^*} (1/\varepsilon^+)(1-y^*)dy^*$$

The energy equation is solved by finite difference calculations after a coordinate transformation from y* to z* to stretch the wall region relative to the central region. The main result of engineering interest is the average total Nusselt number defined by

$$\overline{Nu}_T = \frac{RePr_m}{4(x/D)} \ell n(1/T_b^*) \tag{8-74}$$

where T_b^* is the dimensionless bulk temperature

$$T_b^* = \frac{\int_0^1 T^*(x^*,y^*)u^+(y^*)dy^*}{\int_0^1 u^+(y^*)dy^*} \tag{8-75}$$

and D is hydraulic diameter. An engineering correlation was proposed [19] in the form

$$\overline{Nu}_T \doteq \overline{Nu}_C + \overline{Nu}_R \tag{8-76}$$

where

$$\overline{Nu}_C = 0.020Re^{0.8}Pr_m^{0.33}\bar{F} \tag{8-77}$$

$$\bar{F} = 1+[0.88/(1+0.4(x/D))] \tag{8-78}$$

$$\overline{Nu}_R = \frac{D}{K_m}\sum_{k=1}^{n} \omega_k B_k' A_s^* \bar{\tau}_{WL,k} \tag{8-79}$$

$$\bar{\tau}_{WL,k} \doteq \frac{\ell n(1+t_{D,k}) - \ell n(1+t_{D,k}\bar{r}_{WL})}{\ell n(1+t_{D,k})} \qquad (8\text{-}80)$$

$$\bar{r}_{WL} = \frac{1.37}{Re^{0.37}} \left\{ 1-\exp[-(x/10D)^{0.6}] \right\} \qquad (8\text{-}81)$$

where A_s^* is the slab band absorptance for $\frac{1}{2}\, t_{D,k}$ where $t_{D,k}$ is optical depth based upon hydraulic diameter. The suggestion was made that the results could be applied to either a parallel plate duct or pipe [19].

8.G(c) Sample Application. For example, let us estimate the average heat transfer coefficient due to combined radiation and turbulent convection of steam at a pressure of 60 atm and a reference temperature of 1000 K flowing at 10 m/s inside a 60 mm inside diameter, 1 m long black-walled tube. Nominal property values are assumed as follows: K_m = 0.09 W/m K, $\nu_m = 2.6\text{x}10^{-6}\ m^2/s$, $Pr_m = 0.9$. For simplicity contributions from only the 6.3 μm (k = 1) and 2.7 μm (k = 2) bands will be considered.

We begin by reading values of $\tau_{H,k}$ from Fig. 5-10 for a 1 atm-m PL value. For P = 60 atm and $L_{mbg} = D = 60$ mm we obtain

$$t_{D,1} \doteq (60)(60/1000)(50) = 180$$

$$t_{D,2} \doteq (60)(60/1000)(30) = 100$$

From Table 5-1 the band-width parameters are

$$\omega_1 = 56.4(1000/100)^{\frac{1}{2}}(1.2) = 214\ \mathrm{cm}^{-1}$$

$$\omega_2 = 60(1000/100)^{\frac{1}{2}}(1.2) = 228\ \mathrm{cm}^{-1}$$

Since Eq. (7-66) is derived from spectral integration, the ω values were increased by 20% per the discussion following Eq. (5-58). From Eq. (7-66)

$$A^*_{s,1} = \ell n\,\frac{180}{2} + 0.577 + 0.500 = 5.58$$

$$A^*_{s,2} = \ell n\,\frac{100}{2} + 0.577 + 0.500 = 4.99$$

Next we determine $\bar{r}_{WL}$ and $\bar{\tau}_{WL,k}$. To do so we need Reynolds number,

$$Re = \frac{UD}{\nu_m} = \frac{(10)(60\mathrm{x}10^{-3})}{2.6\mathrm{x}10^{-6}} = 2.3\mathrm{x}10^5$$

From Eq. (8-81)

$$\bar{r}_{WL} = \frac{1.37}{(2.3\mathrm{x}10^5)^{.37}}\left\{1-e^{-(1000/600)^{0.6}}\right\} = 0.011$$

From Eq. (8-80)

$$\bar{\tau}_{WL,1} = \frac{\ell n(1+180)-\ell n(1+1.9)}{\ell n(1+180)} = 0.80$$

$$\bar{\tau}_{WL,2} = \frac{\ell n(1+100)-\ell n(1+1.1)}{\ell n(1+100)} = 0.84$$

Because the values of $\omega_k A^*_s$ are so large, i.e., because the H_2O bands are so wide, we prefer to introduce the internal fraction from Chapter 1. Thus we make the following substitution in Eq. (8-79)

$$\omega_k A^*_s B'_k = 4\sigma T^3 \Delta f_{i,k} \tag{8-82}$$

To find $\Delta f_{i,k}$ in Table 1-1 as a function of $\lambda T = T/\nu$ we need the upper and lower limits on ν for each band. Returning to Table 5-1 we see that the 6.3 μm band is to be centered at 1600 cm^{-1}, and the 2.7 μm band at 3760 cm^{-1}. Thus

$$\nu_{1,u} = \nu_{1,c} + \frac{1}{2}\omega_1 A^*_{s,1} = 1600 + \frac{1}{2}(214)(5.58) = 2197\ cm^{-1}$$

$$\nu_{1,\ell} = \nu_{1,c} - \frac{1}{2}\omega_1 A^*_{s,1} = 1600 - \frac{1}{2}(214)(5.58) = 1003\ cm^{-1}$$

$$\nu_{2,u} = 3760 + \frac{1}{2}(228)(4.99) = 4329\ cm^{-1}$$

$$\nu_{2,\ell} = 3760 - \frac{1}{2}(228)(4.99) = 3191\ cm^{-1}$$

The limits on λT and f_i are thus

Band 1: $\lambda T = 0.455$ to 0.997

$f_i = 0.741$ to 0.965 , $\Delta f_i = 0.224$

Band 2: $\lambda T = 0.231$ to 0.313

$f_i = 0.237$ to 0.478 , $\Delta f_i = 0.241$

Substituting Eq. (8-82) into Eq. (8-79) and entering the numerical values gives

$$\overline{Nu}_R = \frac{60x10^{-3}}{0.09}(4)(5.67x10^{-8})(1000^3)[0.224(0.80) + 0.241(0.84)]$$

$$\overline{Nu}_R = 57.7$$

The convective contribution is given by Eqs. (8.77) and (8-78)

$$\overline{Nu}_C = 0.020(2.3x10^5)^{0.8}(0.9)^{0.33}[1+0.88/7.67]$$

$$\overline{Nu}_C = 419.3$$

Thus the total Nusselt number is 477. To complete the example we find the average heat transfer coefficient

$$\bar{h} = \frac{K_m}{D} Nu_T = \frac{0.09}{60x10^{-3}} (477) = 715 \ W/m^2 \ K$$

REFERENCES TO SECTION 8

I. Thermal System Analysis

1. Buchberg, H., "Electric Analogue Prediction of the Thermal Behavior of an Inhabitable Enclosure", Transactions of the Am. Soc. of Heating and Air-Conditioning Engineers, Vol. 61, pp. 339-386, 1955.

2. Robbins, W. H., "Analysis of the Transient Radiation Heat Transfer of an Uncooled Rocket Engine Operating Outside Earth's Atmosphere", NASA TN D-62, December 1959.

3. Ishimoto, T., and J. T. Bevans, "Temperature Variance in Spacecraft Thermal Analysis", J. Spacecraft and Rockets, Vol.5, pp. 1372-1376, 1968.

II. Combined Surface Radiation and Conduction

4. Bartas, J. G., and W. H. Sellers, "Radiation Fin Effectiveness", J. Heat Transfer, Vol. 82, pp. 73-75, 1960.

5. Sparrow, E. M., and R. D. Cess, Radiation Heat Transfer, Hemisphere/McGraw-Hill, New York, 1978, pp. 180-190.

6. Hrycak, P., and R. E. Helgans, Jr., "Equilibrium Temperature of Long Thin-Walled Cylinders in Space", Chemical Engineering Progress Symposium Series, Vol. 61, pp. 172-178, 1968.

7. Edwards, D. K., "Anisotropic Conduction and Surface Radiation Around a Hollow Cylinder", J. Heat Transfer, Vol. 102, pp. 706-708, 1980.

III. Combined Gas Radiation and Conduction (see also Ref. [4], Ch. 9)

8. Viskanta, R., "Heat Transfer by Conduction and Radiation in Absorbing and Scattering Materials", J. Heat Transfer, Vol. 87, pp. 143-150, 1965. See also Ref. [10].

9. Cess, R. D., "The Interaction of Thermal Radiation with Conduction and Convection Heat Transfer", Advances in Heat Transfer, Vol. 1, pp. 1-50, 1964.

IV. Combined Gas Radiation and Convection (see also Ref. [4], pp. 190-196 and Chapter 10)

10. Viskanta, R., "Interaction of Heat Transfer by Conduction, Convection, and Radiation in a Radiating Fluid", J. Heat Transfer, Vol. 85, pp. 318-328, 1963. See also Advances in Heat Transfer, Vol. 3, pp. 176-251, 1966.

11. Nichols, L. D., "Temperature Profile in the Entrance Region of an Annular Passage Considering the Effects of Turbulent Convection and Radiation", Int. J. Heat Mass Transfer, Vol. 8, pp. 589-607, 1965.

12. deSoto, S., "Coupled Radiation, Conduction, and Convection in Entrance Region Flow", Int. J. Heat Mass Transfer, Vol. 11, pp. 39-53, 1968.

13. Greif, R., and D. M. McEligot, "Influence of Optically Thin Radiation on Heat Transfer in the Thermal Entrance Region of a Narrow Duct", J. Heat Transfer, Vol. 93, pp. 473-475, 1971.

14. Habib, I. S., and R. Greif, "Heat Transfer to a Flowing Nongray Radiating Gas: An Experimental and Theoretical Study", Int. J. Heat Mass Transfer, Vol. 13, pp. 1571-1582, 1970.

15. Pearce, B. E., and A. F. Emery, "Heat Transfer by Thermal Radiation and Laminar Forced Convection to an Absorbing Fluid in the Entry Region of a Pipe", J. Heat Transfer, Vol. 92, pp. 221-230, 1970.

16. Kurosaki, Y., "Heat Transfer by Simultaneous Radiation and Convection in an Absorbing and Emitting Medium in a Flow Between Parallel Plates", Heat Transfer 1970, Vol. III, paper R2.5, Fourth International Heat Transfer Conference, Paris/Versailles, 1970.

17. Edwards, D. K., and A. T. Wassel, "Interpretation of a Turbulent Radiating Gas as a Low-Prandtl-Number Fluid", Letters in Heat and Mass Transfer, Vol. 1, pp. 19-24, 1974.

18. Wassel, A. T., and D. K. Edwards, "Molecular Gas Radiation in a Laminar or Turbulent Pipe Flow", J. Heat Transfer, Vol. 98. pp. 101-107, 1976.

19. Balakrishnan, A., and D. K. Edwards, "Molecular Gas Radiation in the Thermal Entrance Region of a Duct", J. Heat Transfer, Vol. 101, pp. 489-495, 1979.

EXERCISES FOR SECTION 8

1. A room is warmed by a radiant ceiling 4 by 5 m in size well insulated above. Three interior walls and the floor may be assumed to have zero heat flux as measured at their m-surface (Fig. 1-1). It is agreed to model them as a single surface and use $h_c = 1.3\Delta T^{1/3}$ W/m^2 K to calculate the convective heat transfer coefficient. One exterior wall 3 by 5 m has $q_m = U(T_e - T_s)$ where U is 2 W/m^2 K and T_e is the external air temperature $T_e = 10°C$. Convection from the s-surface to the room air also occurs from this wall as from the others, but that from the ceiling is negligible due to stratification near the ceiling. Due to lack of effective sealing, the room air has a volume flow of 3 room volumes per hour leave it, and a compensating mass flow of cold external air flows into the room air mixing with it. It is desired to heat the ceiling at a sufficient rate to maintain the interior walls at 22°C. Find the heat flow required to do so and the temperatures of the room air and the inside surface at the exterior wall. Neglect gaseous radiation.

2. A patio cover is made of 1.25 cm thick painted wood covered with 0.3 cm of black tar paper and is exposed to solar irradiation from above at 1000 W/m^2 and infra-

2. (continued)

red radiation from the sky. The black ground below the patio cover and the ambient air temperature are approximately equal at 30°C. Wind blowing under and over the cover causes h_c to be approximately 6 W/m^2 C. (a) Find the radiosity of the lower surface of the cover. (b) What is the radiosity of the bottom of a black radiation shield spaced 2 cm below the patio? (c) What is the radiosity, if the underside of the patio cover is painted with aluminum paint (ε = 0.22) before the black radiation shield is installed? (d) What is the radiosity of the bottom of the cover, if the top of the cover is painted white and kept clean (α_S = 0.30, ε = 0.85) and the cover is otherwise unchanged from its original configuration? Assume RH = 50% for sky radiation.

3. An "asymptotic calorimeter" is made from a 50 μm thick nickel foil in the form of a disk 5 cm in diameter placed like a drum head over the lip of a copper cup. The foil is blackened on the top, and a thermistor bridge gives a signal of 5 mV per °C of temperature difference between the center of the foil and its edge soldered to the copper. What signal is expected when the edge temperature is 300 K and the device is exposed in 20°C still air to a 400 K black body environment?

4. A space radiator fin is made of aluminum 0.3 mm thick bridging 7.5 cm between heat pipe saddles. The emissivity of the aluminum is 0.80, and the solar absorptivity 0.20. At the saddles the radiator is 330 K. What is the fin effectiveness and the heat radiated per m^2 when the radiator is exposed to 300 W/m^2 of solar irradiation?

5. A 75 mm diameter hollow cylinder is exposed to solar radiation in space. The cylinder is vacuum aluminized on the exterior and banded with white circumferential stripes. The wall is carbon-reinforced phenolic 0.5 mm thick with a thermal conductivity of 40 W/m K. Solar radiation of 1400 W/m^2 is incident at an angle of 60° from the axis of the cylinder. The cylinder spins rapidly about its axis. It is desired to make the bands as coarse as possible without having any point hotter than 40°C or any colder than 10°C. Specify the widths of the stripes and the spacing between them. The following radiation properties pertain:

   Vacuum deposited aluminum: $\alpha_S = 0.10$, $\varepsilon_H = 0.03$

   White inorganic paint: $\alpha_S = 0.13$, $\varepsilon_H = 0.85$

   Carbon-reinforced phenolic: $\varepsilon_H = 0.90$

6. Find the effective emissivity of the opening of a semicircular groove with diffuse wall reflectivity $\rho_w = 0.70$. Compare the exact solution with that from the simple 2-node network.

7. A spinning spherical mirror made of aluminum 20 cm in diameter with a 0.5 mm wall is discharged at 300 K into space from a space vehicle. The mirror has a hemispherical solar absorptivity of 0.13 and a hemispherical emissivity of 0.03. It is exposed to solar radiation of 1380 W/m^2. Assuming the sphere stays reasonably isothermal, how long will it take to warm up to 330 K?

8. The radiator panels built and exposed as described in Exercise 4 are to cool 400 kg/hr of water from 330 K to 280 K. How large an area is needed?

9. Catalyst particles are carried by a gas flowing laminarly inside a 2.5 cm inside diameter tube. The absorption coefficient for the gas particle mixture is 0.04 cm^{-1}. The particles catalyze an exothermic reaction causing a uniform volume heat source $\dot{Q}_V$. How large can $\dot{Q}_V$ be if the maximum gas temperature is not to exceed 1000 K? The walls are at 500 K, and the gas has a thermal conductivity of $0.10(T/T_o)^3$ W/m K where $T_o = 1000$ K. For a first approximation treat the gas as optically thin and assume the gas and particles are in local thermal equilibrium. (a) Find $\dot{Q}_V$ when the walls have an emissivity of 1.0. (b) Discuss how, in the optically thin limit, a lower wall emissivity would affect your answer. (c) If all the particles have the same diameter D_p and the

9. (continued)

absorption cross-section is equal to the geometric cross-section, find how much heat $\dot{Q}$ would be released per particle as a function of particle size when absorption coefficient and $\dot{Q}_V$ are regarded as invariant (obviously the particle loading would have to vary). For what particle sizes would $T_p - T_e$ exceed 10 K where T_p is the particle surface temperature and T_e is the local gas temperature at the center of the pipe?

10. Powdered coal at 300 K and in 300 K air is suddenly exposed to 1200 K black body radiation under optically thin conditions. The reaction kinetics are such that the coal will ignite when its surface temperature reaches 500 K. (a) Why will too-small particles not ignite? What is the smallest D that will ignite as $t \to \infty$. (b) What is the smallest D that will ignite in time t? What is D for t = 10 ms?

11. Repeat the sample calculation in Section 8.G(c) if P = 30 atm. Be careful to adjust ν_m appropriately.